HUNTER ONE

HUNTER ONE
A History of Jet Heritage

Pioneering Britain's Jet
Warbird Movement

Mike Phipp
with Eric Hayward

AMBERLEY

First published 2009

Amberley Publishing
Cirencester Road, Chalford,
Stroud, Gloucestershire, GL6 8PE

www.amberley-books.com

British Library Cataloguing in Publication Data.
A catalogue record for this book is available from the British Library.

isbn 978 1 84868 244 3

Typesetting and Origination by Diagraf (www.diagraf.net)
Printed in Great Britain

Contents

	Foreword	7
	Acknowledgements	8
	Introduction	9
1	Schoolboy Dreams	11
2	Spencer's Air Force	13
3	For Sale – One Hunter	28
4	Hunter One	36
5	Increased Display Flying	52
6	'... and then I get a pilot'	58
7	Retrenchment	59
8	End of Operations?	65
9	Phoenix Rises	72
10	Jet Heritage	75
11	Raising Public Awareness	84
12	Royal Jordanian Air Force and 'Harry'	96
13	Organisational Changes	106
14	New Era at Jet Heritage	116
15	Museum at Last	127
16	Disbandment	135
	Epilogue	137
	Appendix I	139
	Appendix II	140
	Index	158

F-86A

Foreword

History is generously populated with alluringly labelled 'golden eras', associated with every conceivable subject. For those with the love of aviation, the late 1940s and early 1950s was the 'golden era' of jet-powered aircraft development. This was a time when legions of jet aircraft types were designed and manufactured for all the major governments. Apocryphal tales of the early days of their flight-testing abound, along with historic feats such as breaking the sound barrier and attaining world absolute speed records. Larger-than-life personalities such as Chuck Yeager and Neville Duke were in the public eye.

The 1980s was another so-called 'golden era' in the United Kingdom, with the formation of extensive privately owned and privately funded historic aircraft museums. Such institutions were dedicated to the preservation of some of these magnificent post-war aircraft, both jet-powered and piston-engined models. It was both an honour and privilege to have been involved in the formation and operation of Hunter One and Jet Heritage – undoubtedly two of the most prestigious and visionary of those organisations. Both were dedicated to the preservation of flying examples of early military jets and to displaying them at air shows to the public at large.

Often, details of the efforts involved, the endless hours expended and the love cherished upon these aircraft, as well as the hurdles surmounted in putting such magnificent collections together, is lost to posterity. So it is not only fitting but a well deserved tribute, both to the aircraft themselves and those involved, that books such as this are painstakingly researched and put together. Such books provide a historical record of what went before, the like of which, sadly in many cases is never likely to be seen again.

It was a wonderful time, following in a far humbler way the halcyon days of the government-sponsored 'early jet era' itself. In this we led the way, we had the vision, carefully laid the plans, painstakingly found and restored aircraft. We diligently set the standards, laboriously wrote the rules and, through all of this, we, too, shared the dream.

Adrian Gjertsen FRAeS
Former Director of Flight Operations and Chief Pilot
Hunter One Collection and Jet Heritage

Acknowledgements

The idea for this book has been around for many years, during which time information has been gathered together by both Eric Hayward and myself. What has been recorded is something new, so it has not been the case of referring back to other books. However, by way of a cross check, reference to the numerous monthly aviation publications over the years has been of great help. The various editions of Ken Ellis' *Wrecks & Relics* have also proved invaluable.

Reminiscences and stories have been provided by a number of people connected with Hunter One and Jet Heritage over the years. Mention must be made of Bob Coles, Richard Edwards and Peter March, with a special thank you to Adrian Gjertsen for writing the introduction to this book as well as adding his own recollections.

The photographs have come from a number of sources, credits being given in the appropriate place. These include the authors' own photos – MHP or EGH – plus a number collected by Jet Heritage over the years. It has not been possible to establish every person who provided these, and so they are credited to JHL collection. Again, special thanks to Peter March, who provided some of his vast selection of photographs for us to choose from.

Finally, many thanks to my wife for her editorial input. On this occasion she has found the proof-reading less taxing, as, although it is aviation related, she has found the book more of a story than a purely technical work. However, she looks forward to the day when I turn my attention to a novel.

Introduction

Bournemouth Airport had always attracted a wide variety of aircraft, and so the arrival of a red Hunter in the summer of 1981 did not cause much of a stir. It was a couple of months before I bothered to go and see the new resident – from between the slightly opened hangar doors. The arrival of more former military aircraft a couple of years later merited further investigation, resulting in my meeting up with Eric Hayward. Having established I was an enthusiast, and not just another spotter, the affable Eric happily showed me around. Mike Carlton and, later, Adrian Gjertsen were both friendly, but more tied up with the business side, and so they were not to be seen so often. This was the start of many years contact, during which I soon realised that I was witnessing the start of a new era in aircraft operations – the jet warbird. In later years there were the various events leading to the opening of the Jet Heritage Museum.

As the Airport Historian I accumulated details of the Hunter One and, later, Jet Heritage activities at Bournemouth and elsewhere. I felt that they would come in useful at some future stage. So when Eric approached me a couple of years ago with his collection of records and photographs, it was easy to come up with the joint idea of detailing the events in book form.

Eric's fame within the aviation industry led him to becoming known as 'Mr Hunter'. Over the years many pilots and organisations would refer to him to sort out any Hunter problem. In this way he was involved with other operators in later years. For my part, I think all this justifiably resulted in a smile on Eric's face when he saw all the Hunters at the Kemble Hunter Meet of July 2001, as the vast majority had passed through his hands at some stage.

This history of Hunter One and Jet Heritage covers groundbreaking events of a twenty-year period and includes reminiscences to help bring the events more vividly to life in the reader's mind. We hope there will be more than a touch of nostalgia for many.

Mike Phipp

5During the mid-1970s there was only a small and scattered, but enthusiastic, following in the private warbird arena in the UK. There was no coherent group organisation with sufficient financial backing to purchase, overhaul, certify and operate former RAF jet aircraft on the civil register.

1978 – that was the year it all changed. Some surplus Hunter airframes were disposed of by Hawker Siddeley, and that started a movement which gathered enormous momentum over the following ten years.

From existing records and discussions with former employees, enthusiasts, plane spotters and pilots we have gathered facts, data and anecdotes on the work that went into setting up the jet warbird community as we know it today. Hopefully these operations have given great pleasure to many over the years. Remember: Hunter One was the world's first dedicated airworthy jet fighter collection.

I trust the contents will revive happy memories of the many Hunter One and Jet Heritage workers, volunteers and supporters, and even may be an introduction to new enthusiasts to the working of the world of warbirds.

My thanks to some of the crew who made the Hunter One and Jet Heritage experiences happen and enriched our lives but who unfortunately are no longer with us: Bob Brenton, Mike Carlton, Bob Coles, Joh Davies, Ken Hare, Stephan Karwoski, Geoff Roberts, Bob Tarrant, Dougal Craig-Wood.

Thanks to all those who were employed in this happy time in the creation of a unique and successful flying warbird restoration project that led the way for many years. My reminiscences are dedicated to them all.

Eric G Hayward

1

Schoolboy Dreams

Adult achievements often have their roots in childhood dreams, as boys the world over try to emulate their heroes. Mike Carlton dreamt of becoming a fighter pilot; Neville Duke was one of his heroes and Hunter One was to be the culmination of his dreams. The conception of Hunter One can therefore be traced back to the early 1950s.

Squadron Leader Neville Duke was a highly decorated Second World War fighter ace. On leaving the RAF in 1949, he joined Hawker Aircraft at Langley as a test pilot, becoming their Chief Test Pilot in the spring of 1951. On 20 July 1951 he returned to the public eye, when he took the Hunter prototype on its maiden flight from Boscombe Down. During the summer of 1953 he was involved in the development flying of the prototype and early production Hunters from Hawker's new Dunsfold site. His greatest achievement as far as the public was concerned came on 7 September of that year. This was the day he broke the world air speed record, flying the red-painted Hunter prototype WB188 at 727.6mph over the Sussex coastline.

Neville's wartime experiences were well known to aviation enthusiasts, Spencer Flack and Eric Hayward included. His world speed record was recounted by schoolboys everywhere, among them Adrian Gjertsen, Brian Henwood, Stephan Karwowski and Geoff Roberts – all of whom were to play their part in this story. Following Neville Duke was Mike Lithgow, who gained the speed record flying a Swift. Then there were the exploits of Meteor and Sabre pilots in the Korean War. These were the events that further fuelled the schoolboy's early hopes, which were to become reality twenty-five years later.

Hawker rekindled public interest in red Hunters with displays of their demonstrator Hunter Trainer G-APUX in the early 1960s. Bill Bedford's spinning displays were the highlight of the Farnborough Air Show from 1959 to 1962.

In the 1970s the warbird movement in Great Britain had not really begun. At this time the RAF displayed two Hurricanes and four Spitfires at air shows around the country, with the *Red Arrows* providing additional excitement. However, across the Atlantic there was a number of enthusiasts collecting and flying warbirds, including some early jet fighters. These activities came to the notice of Ormond Haydon-Baillie, who was an RAF pilot serving in Canada.

The inspiration for many pilots – Neville Duke flying the Hawker Hunter prototype during the summer of 1951. At this stage, as with any prototype, Sydney Camm's elegant design was uncluttered by underwing tanks, gun pods, air brakes and suchlike. (Hawker Aircraft)

Ormond came from a wealthy family, and this enabled him to join the warbird collectors. On his posting back to the UK in the summer of 1973, he brought with him a Hawker Sea Fury and two Canadair T-33 Silver Stars. These were first shown at the Biggin Hill Air Fair in May 1974 and then at Greenham Common. The T-33s were painted black and displayed as *The Black Knights*. Ormond and his brother Wensley scoured the world for other historic aircraft to restore – these included Spitfires, Tempests and F-86 Sabres. Regrettably, Ormond was killed in a Mustang crash in July 1977, before further aircraft saw light under their wings, and the organisation folded shortly afterwards.

2

Spencer's Air Force

One of the well-known aviation personalities of the 1970s was Spencer Flack. Although he was never directly involved with Hunter One, without him Hunter One would never have come into being, let alone Jet Heritage. Spencer, an aviation enthusiast, owned a plant hire company and was Chairman of the Elstree Aero Club. He believed in living life to the full. As well as having a private pilot's licence, he had his own hangar and aircraft at Elstree Aerodrome in Hertfordshire. Desiring more of a thrill from flying, he became interested in the idea of restoring former military aircraft. This led to the formation of his 'Elstree Air Force', which included a Bucker Jungmann and Jungmeister trainer. Moving up a stage further, Spencer acquired former a target-tug two-seat Hawker Sea Fury, G-BCOW, from warbird dealer Doug Arnold, during February 1977. Spencer operated this from Elstree as the flagship of his 'air force'. Early in 1978 he heard of another former museum Sea Fury that was up for sale, and decided to add it to his collection. However, Spencer still had one further goal – to add a jet fighter rating to his civilian pilot's licence, thereby achieving a life-long ambition.

Around this time Spencer had established a contact within British Aerospace's Design Office (formerly Hawker's) at Kingston-upon-Thames. Aviation Designer, Bob Cole.

Bob: I frequently accompanied Spencer on G-BCOW's outings – it seemed a shame to waste the back seat! Spencer mentioned that he would like to be the first to operate, privately, a military jet fighter, but nothing much had transpired. One trip was to BAe's 1977 Families Day at Dunsfold Aerodrome, and this was the beginning of the Hunter saga. As we taxied in, we passed a line of picketed former Danish Air Force Hunters, and Spencer's head was turned towards them. He wanted to know what they were doing there, and I explained that they were now redundant. Spencer replied "Can you get hold of one? I think it's time we joined the Jet Set".

So, on the following Monday I made enquiries, but was told the airframes had already been allocated. However, E-418 had been set aside for the local Surrey & Sussex Aircraft Preservation Group, but for some reason they had been unable to collect it. Later I was able to make contact with the group, and found they were prepared to pass ownership to me. The wheels were set in motion.

Left: Spencer Flack with his two-seat Sea Fury G-BCOW in the summer of 1978. The wording by the cockpit reads 'Capt. S.R. Flack – Elstree Air Force'. The delight of owning and flying this aircraft motivated Spencer to buy a further Sea Fury and a Hunter. (British Aerospace)

Below: Spencer was not happy until he acquired a jet fighter for his air force, in the shape of a former Danish Air Force Hunter. Seen shortly after arrival at Elstree in May 1978, E-418 would need a lot of tender loving care before returning to the air. (EGH collection)

E-418 had originally been delivered by Hawker's to the Danish Air Force as a Hunter F.51 in June 1956. After almost twenty years service, it was withdrawn in 1975 and placed into storage at Aalborg Air Base. It was one of eighteen surveyed by BAe during that autumn, following which in December 1975 the whole batch were purchased, dismantled and returned to Dunsfold by road. Allocated B-class registration G-9-440 by BAe, E-418 was held in store with the others for possible refurbishment and resale. However, with production development of the Harrier and the new Hawk under way, there was no capacity at Dunsfold to handle these Hunters, which remained in external storage while being cannibalised for spare parts.

After Bob's negotiations, some months passed before E-418 was ready to depart Dunsfold by lorry for Elstree – ostensibly for display at Spencer's 'Elstree Air Force' Museum. BAe had placed no restrictions on the use of the airframe, as all surplus Hunters were normally donated to museums for static display only. Had it been realised at the time that it was to be restored to flying condition, BAe probably would not have agreed to Spencer taking ownership. It arrived at Elstree in May 1978, having been preceded by former Royal Navy Sea Fury FB.11 WJ244, which arrived by road from Ashford (it had been at Southend Museum before) in April. In July 1978 Spencer appropriately registered the Sea Fury as G-FURY and the Hunter as G-HUNT. Not content with this, the following January he purchased Spitfire FR.XIV NH904 from the Strathallan Collection.

Bob: Spencer now had his jet fighter, but it could not be certified for flight by the Civil Aviation Authority (CAA), let alone registered, without serious and experienced help from someone with a full working knowledge of Hunters. I knew of Eric Hayward, who was a Senior Airframe Inspector with BAe at Dunsfold. He had twenty years' experience working on Hunters, and was regarded as quite an authority on the type. I contacted Eric on behalf of Spencer to ask whether a Hunter could be restored and placed on the Civil Register by a private owner. If so, would he supervise its restoration? Eric's immediate response was "No – the CAA will never agree".

However, I considered the phone call as one of the most significant I ever made, as it was carefully planned and tailored to work like a double-edged sword. It would introduce Eric to the project, with the opportunity to 'come on board', and gave me a rare chance to put my psychology course to the test. I felt sure that Eric would show his hand. Being the total aviation professional that he is – not only on Hunter matters – I knew it would be difficult for him not to want to follow the subsequent proceedings.

In fact Eric had already been acquainted with Spencer's Hunter, as he was the engineer sent to Aalborg to survey the stored Danish Hunters. It was on his recommendation that BAe brought the aircraft back to Dunsfold for possible rebuild. At this stage it was not realised what an important part Eric would play in later developments of the jet warbird movement.

Left: Two Hunter experts were soon 'recruited' by Spencer to help with the restoration of E-418. Bob Cole was from the Kingston Design Office and is seen here with Eric Hayward (seated in aircraft), who was a Senior Airframe Inspector at Dunsfold. (British Aerospace)

Below: The two recruits were given flights by Spencer in his two-seater Sea Fury at the British Aerospace Families Day at Dunsfold in June 1978. Spencer and Eric are seen standing alongside G-BCOW after Eric's 'flight of persuasion'. (EGH Collection)

Eric: Each year BAe held a Families Day, which enabled the latest Harriers and Hawks to be shown off, along with various visiting aircraft. For the June 1978 display Spencer Flack again brought down his two-seat Sea Fury, and I was introduced to him by Bob before the display. He said that he had been given my name by Bob as an expert on Hunters. He explained that he had just acquired a Hunter hulk which he was going to restore to an airworthy condition – would I help him? The request came out of the blue, as I thought he had given up the idea. At the time there was no way I could see a Hunter flying in civilian hands, so I again replied 'No'. Spencer seemed rather taken aback by this, and said he would change my mind. After his flying display he came up to me again, exploiting the virtues of being able to fly a fighter around the sky. I still thought him rather mad and again said 'No'. This was not the reply Spencer wanted. He then asked if I was used to flying a fighter, and he was surprised to learn that I had never flown in one. 'Here's your chance' he exclaimed, 'come for a flight in my Sea Fury and I'll give you a demonstration'. There followed an exhilarating twenty minutes over the Surrey countryside, which resulted in my being able to see his point of view, So when he asked again 'Now will you help me build my Hunter?', having been hooked on the project I replied 'OK, but there's no guarantee the CAA will allow it'.

Events at the Families Day changed Eric's life forever and possibly started the jet warbird industry as we now know it, leading to an era where the CAA agreed to the licensing of former military jets, which are now in civilian hands all over the world.

Spencer's newly acquired Sea Fury and Hunter had both been standing in the open for a number of years. On arrival at Elstree they were found to be in reasonable condition, with little corrosion, and most instruments were in place in the Sea Fury. However, the Hunter had been cannibalised, it had no engine and was missing many instruments and systems. Restoration of the airframes was undertaken in two separate hangars, with work lasting around two years on each aircraft. During this time Eric made many visits to Elstree to work on the Hunter, in company with Norman de Viell (a colleague from BAe), Bill Final and others. From October 1979 Spencer would send one of his aircraft to Fairoaks to collect the team of restorers, to save a road journey. The team had access to technical drawings and documents, with Spencer providing the funding. First they had to strip the airframe down fully, before the rebuild could commence with spare parts found from various sources. These included two of the other former Danish aircraft, E-423 and ET-272, arriving by road from Hatfield to yield parts for the rebuild. Then a replacement engine had to be found.

Bob: Half jokingly Spencer asked if I happened to have an Avon engine in my garage. I didn't, but one soon came to light. News of the restoration soon spread around the Design Office at Kingston, where a staff member approached me to say he was an officer with Balham ATC Squadron, and the cadets had an Avon for instructional purposes. That evening we went to have a look, and I was delighted to find it was the correct mark for a Mk.51 Hunter. Also, it had been very little used,

Work under way on E-418 at Elstree during 1979, with Eric test-fitting the engine. The Avon engine was a Rolls-Royce overhauled one, rated at zero hours and acquired by Spencer from a London ATC Squadron. (P Boyden)

The Hunter was complete by February 1980 and is seen unpainted at Elstree prior to its first engine runs. At the time Eric was concerned at a jet fighter taking off from Elstree's runway, but Spencer knew he had a capable pilot on board the project – Stephan Karwowski. (EGH collection)

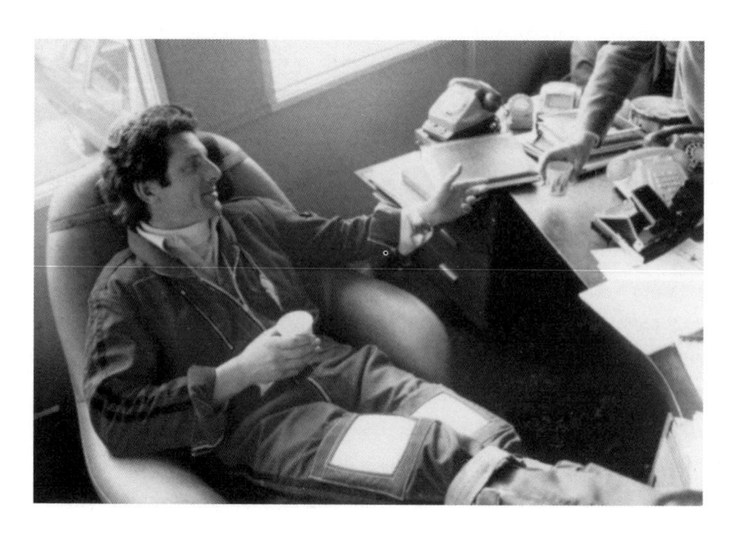

Stephan at rest in the crew room at Elstree. He was already an experienced Hunter pilot, and worked with the restoration project most of the time. More than happy to undertake the early flight trials of G-HUNT, he later undertook the early flights of Brian Kay's G-BOOM. (P Boyden)

and, apart from some toffee wrappers stuffed up the intakes and outlets, it looked in remarkable condition. Rolls-Royce checked the paperwork and confirmed that they had serviced it back to zero hours before it became surplus to requirements. This was good news for Spencer, who then negotiated its acquisition with the Squadron. He suggested that in exchange for the engine he would give the cadets a day out at Elstree, with a flight in his King Air thrown in – unanimous agreement!

Rolls-Royce engineers surveyed the Avon's serviceability, with specialists from British Airways at Heathrow checking it and conducting the initial ground runs on a test bed at the end of February 1980. G-HUNT completed all functions and tests successfully, including having a new electric starting system fitted, and by March was looking like a complete aircraft again. Its significance was that it was the first jet fighter to be rebuilt in the UK from a hulk. Another group of enthusiasts had flown Vampire Trainer WZ507 in February, but that had always been maintained in reasonable condition since being demobbed from the RAF.

Eric: At Dunsfold I had been used to seeing Ministry of Supply Maker's Certificates confirming Hunters had been built to necessary standards, but obtaining a civilian C of A was completely different. I knew one of the local CAA Design Surveyors and he gave me the name of Mr Clifton, the CAA's Southern Areas Chief. Having made contact, I expected to be summoned to his office in London or Gatwick, but was called to his home near Camberley on a Saturday morning. He seemed affable, and suggested we talk in his garden shed over a couple of beers. We seemed to talk about the state of the aircraft industry, people we knew and other irrelevant matters. I can't recall he asked me much about my experience in Hunter matters – at least that's how it seemed at the time. Eventually he said, "That's fine then, contact our Luton surveyor, Colin Turner. I will tell him we've spoken and see what you can both set up. And good luck with your project!" I left on Cloud Nine!

Colin Turner proved to be an excellent surveyor who guided the team whilst also demanding an extremely high standard of work. Over the months the restoration team earned Colin's confidence, which resulted in him eventually signing, without hesitation, G-HUNT's Permit to Fly (Test-flying) on 20 March 1980. This had initially posed a problem to the CAA, as a Permit to Fly mainly applied to home-built aircraft. It had been introduced in 1935 (as Authorisation to Fly) to cover the various ultra-light aircraft, such as Flying Fleas, that were appearing at the time. It was never intended to cover a jet fighter, but there was no other category in which to place G-HUNT.

Eric: I asked Spencer where he intended to fly from for the maiden flight, knowing that Elstree's 2,000ft runway was not ideal for such a flight, and that Hatfield would be more suitable. "We'll fly it from here!" was his reply. "I know a pilot (Stephan Karwowski) who can fly it out when the conditions are right." I agreed, on condition that the

> temperature and wind direction were correct, and that the Hunter carried minimal fuel and no underwing tanks. Satisfactory engine runs were undertaken early in March, and an afternoon later in the month provided the right conditions for Stephan.

Stephan was a former RAF pilot who had undertaken some excellent Hunter display flying. Already known to Spencer, Stephan was well qualified to undertake the flight. With dusk falling over Elstree on a cold 20 March, the unpainted Hunter was pushed by Range Rover to the very end of the runway in order to get the maximum take-off length. Fuel was only carried in the main fuselage tanks, there were no drop tanks or pylons fitted at the time. Stephan opened up, and when he came into view over the hill on the runway he was already committed, as at its end was a belt of oak trees and a school. Full power enabled G-HUNT to be airborne well within the runway distance, with Stephan returning for an unplanned fly-past and a victory roll. He then landed the Hunter at Cranfield, from where further test-flying was undertaken. Stephan was flown back to Elstree in Spencer's Twin Comanche. Later that night toasts were drunk and the drinks flowed freely. On the wave of success all got overly merry, the alcohol rushing to their heads after a long and tiring day: 20 March 1980 – the day that G-HUNT was born.

The restoration team tried to claim that G-HUNT was the first jet to fly from Elstree, but it turned out that it had already been used by a Citation executive jet. However, they were able to claim other firsts for G-HUNT: first privately owned jet fighter on the British civil register; first privately owned aircraft on the British register fitted with ejector seat and fastest privately owned aircraft on the British civil register at the time

Ken Ellis, the Editor of *Wrecks & Relics*, reported G-HUNT's restoration as a ground-breaking event. On 21 March Stephan flew G-HUNT for a photo shoot

Hunter G-HUNT on its maiden flight from Elstree on 20 March 1980, seen from Brian Kay's Learjet. The front nose wheel door is not fully closed, but this did not deter Stephan from continuing the planned flight to Cranfield. (EGH collection)

undertaken from Brian Kay's Learjet G-ZOOM, and then landed at Leavesden for the Hunter to be painted in a red scheme finish by HPB Spraying. Then it was back to Cranfield for additional test-flying, prior to G-HUNT's first public display at the Biggin Hill Air Fair on 17 May. Spencer undertook his desired flight from Cranfield on 28 July, having been delayed while awaiting CAA paperwork authorising his licence amendment. Initially he took off without the flying controls switched to 'power', as Stephan had overlooked this in the rushed pre-flight briefing. Spencer complained about the sluggish controls, and immediately landed for the error to be corrected. He then took off for a second time, flying incident-free to Bedford before returning. Spencer's Seneca acted as chase plane. At it turned out this was the only time that Spencer flew his Hunter, and Eric still had his doubts about Spencer actually having CAA authority to fly the aircraft. Whatever the case, Spencer now owned a potent machine, capable, on paper, of 620kts/Mach 1.3 and a ceiling of 55,000 ft (civil airliners normally fly at 35/40,000 ft).

With G-HUNT flying, the team were then able to concentrate their efforts on the Sea Fury, being rewarded in June 1980 when G-FURY also took to the air again. They then turned their attention to Spencer's Spitfire, which was registered as G-FIRE in March 1979, having just been beaten to the mark G-SPIT by fellow collector Doug Arnold. It made its first flight on 21 March 1981, piloted by Ray Hanna, flying to Leavesden for painting. Spencer then started work on the construction of a Christen Eagle aerobatic biplane from a kit. To part-finance the restorations, he sold his original Sea Fury G-BCOW to a Californian owner in July 1980 (it was to return to the Royal Navy Historic Flight in 2007).

To help complete the story leading up to Hunter One, mention needs to be made of two other pilots. First was Adrian Gjertsen, an RAF Gnat & Hunter QFI who left the service in 1978, undertaking 'civilian' training at Leavesden prior to joining Britannia Airways as a Boeing 737 captain.

Spencer's aim was to be able to pilot a jet fighter. Assisted by Eric Hayward, he is seen completing the necessary CAA paperwork prior to undertaking his first flight in G-HUNT, from Cranfield on 28 July 1980. It was to be Spencer's only flight in G-HUNT. (EGH collection)

Adrian: Whilst at Leavesden I heard about this madman who had bought a Hunter which he planned to fly. My curiosity was aroused, so I drove over to Elstree, where I found Hunter bits outside a hangar, with the fuselage on stands inside. Taking a closer look, I was spotted by an officious-looking gentleman who asked what I was up to. I explained that I had recently finished flying Hunters in the RAF, and knew some technicians who might be able to help with the restoration. The fellow said he would give my phone number to Mr Flack and ushered me out of the hangar.

Two weeks later the phone rang, and it was Spencer asking me over to Elstree. On arrival at the hangar I was directed to a small office, where I found the officious gentleman, who asked if I remembered him. Of course, it was Spencer! He was a tremendous character who, before I left, promised me a chance to fly G-HUNT upon its completion.

Although I never expected to hear any more, true to his word Spencer contacted me a couple of years after our meeting. He offered me the chance to fly G-HUNT as back-up to Stephan – no need to guess my reply! So, in July 1980, I was given the opportunity to display it at Dunsfold's Families Day. Although I was not aware of it at the time, this was before Spencer had flown G-HUNT himself.

A second potential civilian Hunter owner now appeared in the shape of Brian Kay of Ambrion Aviation, which was based at Leavesden. Brian was another London businessman who, as already mentioned, flew a Learjet from Elstree. Knowing what Spencer was up to, Brian, through the efforts of Bob Cole, acquired surplus Hunter Trainer ET-274 from BAe at Hatfield in June 1979 (plus ET-272 from Spencer for possible further spares). This turned out to be another of the former Danish Hunters which Eric had surveyed at Aalborg back in 1975. Brian had the Hunter moved to Leavesden where, led by Eric, restoration was undertaken by almost the same team as for G-HUNT. As before, they were collected from

G-HUNT was painted in Spencer's trade-mark red colour scheme at Leavesden and appeared at a number of air displays during the summer of 1980. The ground crew referred to it as 'Spencer's Triumph TR.7'. (EGH collection)

Shortly after Spencer acquired G-HUNT, a second former Danish Air Force Hunter was bought by Brian Kay. It was rebuilt at Leavesden during 1980-81 as G-BOOM. First flown by Stephan in April 1981, it was based at Stansted, also wearing a non-military colour scheme. (JHL collection)

Spencer's trio of fighters in action – Sea Fury G-FURY, Hunter G-HUNT and Spitfire G-FIRE – July 1981. This was shortly before Spencer's near-fatal crash in the Sea Fury at Waddington, which resulted in his putting the rest of his air force up for sale. (JHL collection)

Having departed on delivery to Denmark in June 1956, G-HUNT returned to its birthplace at Dunsfold for the Families Day in July 1980. Seen after its display are the well-pleased team of Spencer, Adrian, Bob and Eric. (EGH collection)

Fairoaks and flown up to Leavesden. Restoration was completed by early 1981 and the Hunter, now registered G-BOOM, made its first flight on 25 April, piloted by Stephan. A second flight was undertaken to Stansted on 3 June, with Brian Kay then making his first ever Hunter flight later the same day.

For his 'Air Force', Spencer choose an overall red colour scheme, set off by a blue and white strip along the fuselage and across the fin and rudder. Purists were horrified that a Spitfire should appear in such a scheme, but Spencer wasn't worried. Although the Sea Fury and Spitfire could operate from Elstree, the short runway there meant that G-HUNT had to be housed at Cranfield. The aircraft participated in a number of displays during 1980 and 1981, with Spencer flying the Sea Fury or Spitfire, assisted by Pete Shepherd (former RNHF pilot) on the Sea Fury, and with Stephan flying G-HUNT. Eric and the rest of the 'volunteer' engineers continued to keep an eye on the aircraft. As mentioned, G-HUNT's first public display was at the Biggin Hill Air Fair on 17 May 1980, followed by Dunsfold's Families Day on 10 July. Other displays during August were Alconbury, Dublin, Munster and Yeovilton. Later in the month underwing drop tanks were fitted, which gave G-HUNT the range for a display in Belgium, followed by Rotterdam's Anniversary Display on 13 September in company with G-FURY. So ended G-HUNT's first operational summer.

> *ERIC:* The display of G-HUNT at Dunsfold caused some problems. The attitude of BAe on hearing that one of their employees had been involved in restoring to flight one of the Hunters they had officially scrapped was very hostile. The employee concerned was warned that he jeopardised his position in the company if he persisted. A threat that he ignored as he intended to leave the company at the end of the year anyway. G-HUNT returned to its birthplace in June on the invitation of the Dunsfold test pilots for an inspection and short display. So there was little BAe could now do to block the project. Yes, that employee was me! A further complication was that Stephan had just left BAe's employment after a disagreement. Hence the reason for Adrian flying G-HUNT on the day. I left BAe in December and moved away to work for Airwork near Bournemouth.

There were a variety of shows for Spencer's 'Air Force' during the spring and summer of 1981, including Biggin Hill and Mildenhall Air Fete in May, Dunsfold's Families Day, Donnington Park, Liège and Tollerton in June, with Lossiemouth in July. During July Rod Dean – another experienced RAF Hunter pilot – was cleared to fly G-HUNT, displaying it at Yeovilton. For his return flight to Stansted on 20 July Rod made a touch-and-go at Boscombe Down, prior to overflying Tangmere and Dunsfold. This was because he was carrying First Day Stamp Covers to commemorate the thirtieth anniversary of the Hunter's first flight. Spencer's intention was that all three fighters would fly together as his 'Air Force', but this plan did not last as long as Spencer had hoped. On 1 August Stephan and Pete flew G-HUNT and G-FIRE to Yeovilton's Air Day, while Spencer took G-FURY to Humberside's Show. Flying over Lincolnshire on 2 August whilst returning to Duxford, the Sea Fury suffered a major engine failure following a loss of oil pressure. Spencer headed for nearby Waddington but failed to make the runway. Coming down in the undershoot area, Spencer was lucky to get away with his life as the Sea Fury broke up and burst into flames. Rushed to hospital he remained in intensive care for three weeks. After 3½ months he returned home on crutches for a long recuperation.

Although Spencer was in hospital, plans were already in place for other air show appearances. During August G-HUNT was displayed by Stephan at Antwerp, Fairyhouse Racecourse (north of Dublin), Leicester and Staverton. Adrian displayed it at Valley's Open Day on 15 August, where the Hunter suffered a radio failure requiring Eric and his team to fly up to carry out repairs. There were frequent radio frequency problems with the aircraft during the early months of operations due to the incompatibility between military and civil radios.

Adrian: I flew HUNT from Valley to Coventry, where I displayed it at the RAFA Show on 16 August. It was still an unusual and exhilarating experience to be flying a civilian Hunter. On landing after my display HUNT suffered a brake failure and ran onto the grass over-run area. The fire section was soon on hand, although there

G-HUNT stuck in the grass at Coventry. Fire crews ponder how to return the Hunter to the runway following its brake failure at the RAFA Air Display in August 1981. Air bags under the wings eventually solved the problem. (A Gjertsen)

Spencer's Sea Fury …

… and Spitfire.

appeared to be no damage. When I jumped down from the cockpit I could see the Hunter was sinking into the damp ground, and this meant that it could not easily be moved back to the runway. The fire chief's answer was to bring in some old railway sleepers under the aircraft as a base and then use air bags to raise it out of the mud. Further planks were placed under the wheels and ropes were attached to the undercarriage legs to tow it back to terra firma. Whilst this was going on we all had to cope with the deafening noise of a British Airways Concorde which was magnificently making low passes overhead during its part in the show. (One of the very early public Concorde displays). Obviously I had to tell someone of the over-run incident. Spencer was in no condition to receive such news and so I contacted Eric. As HUNT could not be flown out that day I made my own way home.

Eric: I received a call at home from Adrian who explained the problem, adding that Coventry wanted the Hunter off their airfield a.s.a.p. So, the following morning I set off in one of Airwork's vans, my boss having given the OK. On board were a spare main wheel, my toolbox, a tin of hydraulic fluid and a lot of spare brake hydraulic pipes. I arrived at Coventry to find G-HUNT beside the control tower, with the starboard wheel spat badly damaged, and various hydraulic pipes having suffered grievous injuries. On talking to a fireman who had assisted the recovery, it seemed that most of the damage had occurred when recovering the aircraft back onto the runway. So, with the help of the fireman, I went to a nearby hangar and borrowed some jacks to jack the aircraft up. I removed the damaged spat, cleaned off the mud and grass, replaced and repaired the necessary pipes, refilled the hydraulic and bled the system – a good day's work I thought.

The next morning I carried out the necessary function checks, removed the jacks and considered how to test my work. Up until this time I had only ever started a Hunter a few times, and certainly never taxied it. Now, with a few spectators around waiting to see if the brakes would work, I actually drove it like a would-be fighter pilot. Adopting my most casual and efficient air, I checked the aircraft, contacted the tower for permission to start up and taxi. This was agreed, all went well and, as I parked the aircraft and shut down, one of the bystanders shouted "God, how I envy you professionals". Little did they know!

I left Coventry later that evening in the van, with Stephan collecting G-HUNT the following day, and flying back home minus a wheel spat. In later years, whenever I started a Hunter and taxied it, I always thought back to that first time.

September saw continental visits to Münster and Soesterberg Open Day, after which G-FIRE and G-HUNT finished their 1981 display commitments.

3

For Sale – One Hunter

As Spencer slowly recovered from the Waddington crash, his wife said it was time he reassessed his future. Not wanting to risk a similar incident, he decided to sell his Jungmeister, Hunter and Spitfire. So, in the summer of 1981, he spread the word around his aeronautical contacts to see if there was any interest. This brought out Mike Carlton, who expressed his interest in taking on the Hunter. Spencer later undertook more sedate flying, establishing five FAI records in one day in October 1983 flying a Piper Seneca between the four capital cities of the UK. Then in 1990 he won the Schneider Race in his red Beech Baron G-FLAK. Sadly, Spencer was killed in Australia in February 2002 participating in his other passion – historic racing cars.

Mike Carlton was a young, wealthy London businessman, the Chairman and Managing Director of a number of property companies, including the Brencham Group. He also had interests in property in Scotland. As a teenager he had learnt to fly gliders with the ATC, but his application to join the RAF as a pilot was turned down due to 'a lack of aptitude for flying'! He later went on to obtain his PPL on Tiger Moths at Biggin Hill. In aviation circles he was a well-known glider pilot, the holder of nine British National Records in the late 1970s and early 1980s. He also captained the British team in the World Gliding Championships held in Germany in 1981. Mike became interested in the aircraft preservation scene, becoming a member of the Historic Aircraft Association. However, he wanted something different to what other owners were flying at the time. So, a meeting was set up with Spencer, following which he agreed to sell G-HUNT to Mike on the condition that he continued to operate it as a display aircraft. Mike inspected the aircraft at Stansted and the deal was agreed in September 1981, with Mike officially becoming the owner of G-HUNT in October. So, Mike now owned a civilian fighter, but where to keep it and who was going to look after it? He had contacts at Biggin Hill, but were they dedicated enough? Spencer had the answer.

Spencer told Mike that Eric Hayward was his man, as without his efforts at Elstree the Hunter would never have taken to the sky in the first place. By now Eric had left BAe at Dunsfold and had moved to the South Coast, where he was working for Airwork at their Ferndown site as Technical Officer, Hunter Spares. Like Mike he had been turned down by the RAF as being unfit. Luckily this had not deterred him from pursuing a career in aviation.

Eric: I was contacted by Mike Carlton to see if I would become his 'spare time' engineer to keep G-HUNT flying. Initially I said no, as I had just moved home to Ringwood, and considered Stansted was on the wrong side of London for travelling. Mike said, "No problem, I'll bring the aircraft to you!" He found accommodation in Glos-Air's hangar at Bournemouth and arranged to have G-HUNT flown down in September. Due to my affection for Hunters built up over twenty-seven years with Hawkers, my motto was "My work is my hobby. My hobby is my work". Knowing this, Mike realised that there was no way I could say "No", so I was hooked for a second time!

Glos-Air Services had been established in Hangar 600 at Bournemouth Airport in the summer of 1973, specialising in overhaul work on Islanders/Trislanders, the Aero Commander range, plus maintaining business executive twins such as Jetstreams and King Airs. A fighter was something new! Bournemouth proved suitable for fast jet operations as there was plenty of empty airspace in the area, unlike Stansted or Biggin Hill. Limited scheduled services were operated by Dan-Air, whilst on the far side of the airport One-Eleven production was just coming to an end. At the time the airport was also hosting successful air pageants each summer.

Adrian: Following his hospitalisation, Spencer phoned me to say he was selling G-HUNT to Mike. I wrote to introduce myself, explaining that I had flown for Spencer. Mike later got in touch with me and asked me to fly the Hunter, the only proviso being that I also taught him to fly it. Mike had fallen out with Stephan, who was well known for doing things his own way (he eventually left in the spring of 1982). So, in the autumn of 1981, began an association and friendship with Mike and Eric, plus the start of frequent visits to Bournemouth.

Following purchase by Mike Carlton in September 1981, G-HUNT moved to Bournemouth to be maintained in the Glos-Air hangars. This was under the eye of Eric Hayward, who had also moved to the Bournemouth area and so continued his connection with G-HUNT. (EGH coll)

Citation G-UESS was operated by Trans European Air Charter, one of the companies with which Mike Carlton was associated. It undertook charter flights, mainly from Biggin Hill. Note the Brencham Group badge on the fin. (Image in Industry)

Trans European Air Charter also operated Harvard G-TEAC which was restored into RAF colours as MC280. Again based at Biggin Hill, it served mainly as a company 'hack' and is seen in company with G-HUNT. (EGH collection)

At the end of 1981 G-HUNT was broken down at Bournemouth so that its first annual overhaul could be undertaken. It was during this time that Mike Carlton decided that the paint scheme should be modified before the following display season. (EGH collection)

Mike already had connections with Bournemouth Airport. One of his companies, Osiwel Ltd, purchased a Citation executive jet which was overhauled by Glos-Air during May 1981, emerging in June as G-UESS. Mike also flew a Harvard trainer registered G-TEAC. This represented TransEuropean Air Charter – one of Mike's Biggin Hill companies. Later flown in RAF silver training markings with the serial MC280 (Mike Carlton & part of its former RAF serial), it was frequently used as a company hack, visiting Bournemouth on a number of occasions.

During the winter months of 1981 G-HUNT was broken down for overhaul prior to renewal of its Permit to Fly, with attention given to improving its electrical starting system. Discussion with Eric led to the decision being taken to have the Hunter repainted overall red – deleting its stripe lines in the process – so resembling Neville Duke's record-breaking Hunter of 1953. The spring of 1982 saw G-HUNT test flown again, departing on 9 April for the repaint at Little Staughton.

> *Eric:* Little Staughton was virtually derelict, the Hunter landing OK on the former main runway which was strewn with small stones and debris. It was then towed to one of the old hangars, which was now a paint shop, and it just fitted in. We then discovered there were no services or fuel. We had flown in light for safety's sake, but when we came to recover the Hunter we had to bring our own fuel. So we arrived with three 40-gallon drums and a hand pump in the back of our Ford van. Carefully climbing on the newly painted centre fuselage, we gravity-fed the fuel into the tanks. There was a general air of urgency to get the aircraft out – I never really did get to the bottom of it, but suspected the airfield was closed and we should not have been there at all. The Hunter was towed out onto the runway, which we had swept to clear a path for takeoff. Stephan lifted off without flaps, swiftly retracted the undercarriage and flew over us at about 15-20 feet going flat out. This left us observers on the ground in a storm of dust and debris, with Stephan being well on his way to Stansted to top up with fuel. As ever, in his own style, he would fly "anything, anywhere".

Mike realised he had a problem. Although he had long held a pilot's licence, it did not cover jet fighters. So, although he owned G-HUNT, he was going to have to rely on Stephan or Adrian to fly it for him, which was rather stupid. Adrian had been a Hunter QFI at RAF Valley. So, on his suggestion, Mike approached the RAF late in 1981 to see if they would give him a conversion course at Valley onto Hunters. The RAF did not want to get involved with civilians, although they provided him with ground school and simulator training at Brawdy. But they did not want to provide flying, quoting a price of £8,000 an hour for the eight to ten-hour training programme. Even for a businessman like Mike this was a ridiculous price at the time, and so he did not proceed. What could he do now?

To backtrack a little, as mentioned, Eric had also been working on Brian

G-HUNT's revised paint scheme was overall red, without the blue and white strips. Seen outside the Glos-Air hangar at Bournemouth in spring 1982, at this stage its registration on the tail is all white. This was soon amended to be infilled with black. (MHP)

Above left and right: Brencham Group's boss and Hunter owner Mike Carlton, seen in company (left) with his 'spare-time' engineer Eric Hayward. On the right Mike is seen with Adrian Gjertsen, who was his Hunter instructor during 1981, as well as team display pilot. (EGH collection / Peter R March)

Kay's Hunter at Leavesden. Again, Spencer had pointed Brian in Eric's direction for help on the rebuild, resulting in G-BOOM taking to the air in April 1981. However, little flying was undertaken by Brian in his new acquisition. This enabled Mike to come up with the answer to his training requirement. He knew a Hunter QFI and he knew of a Hunter Trainer. Now to bring the two together. So, early in 1982 Adrian and G-BOOM appeared at Bournemouth so that Mike could work out a deal. This resulted in Adrian giving Mike the necessary conversion training on G-BOOM during the spring and summer, to enable the CAA to issue the required endorsement for his licence. From October Mike was a certified jet fighter pilot! Mike first flew G-HUNT on 11 November – his dream achieved! Had he been able to wait a while longer, Adrian later obtained CAA approval for Type Rating as a Hunter examiner.

Amongst the display work in 1982 G-HUNT appeared at Calshot on 26 June, Humberside on 4 July, Lee-on-Solent on 17 July, followed by the Bournemouth Air Show on 18 July, Brawdy Air Day on 22 July and Yeovilton's Air Day on 31 July. The next month took in Alconbury's Air Tattoo on 14 August and Fairyhouse Racecourse on 22 August. There was an overseas trip to Toulouse from 18 to 20 June where the Hunter suffered another radio failure. Again this required Eric to fly out by Navajo to rectify the problem. It struck Mike that G-BOOM would have an advantage in being a two-seat aircraft which could carry an engineer on away trips. Although he had a Harvard, he needed something faster. Halfway through his Hunter conversion training, Mike spoke with Brian Kay to see if he would be willing to sell G-BOOM, but Brian was initially asking too high a price. However, by the end of Mike's training sorties, the two owners agreed on a price, resulting in Mike becoming the owner of a second Hunter in October. At the same time the hulk of ET-272 was also acquired to provide spare parts. Brian had G-BOOM painted in a pleasing dark blue and light-grey colour scheme, based on the prototype Northrop F-20 Tigershark. Mike felt this looked out of place next to G-HUNT, and so he had the two-seater resprayed during the spring of 1983 into a matching red colour scheme. It could now be used as an airshow performer, as well as a training and support aircraft. Harvard G-TEAC was then sold at auction at Duxford in April.

By now Mike had well and truly caught the fast jet bug and was looking for other former RAF jet aircraft to purchase. His first choice was simple – the Jet Provost primary trainer – but he went a long way to find them. In the spring of 1983 the RAF had no suitable surplus aircraft available, but Eric found a source. The Royal Singapore Air Force had advertised its fleet of surplus Jet Provosts for sale. His enquiries met with success; he was advised that the aircraft were well maintained and had low-time airframes. So, Mike arranged to fly out to Singapore with Eric at the beginning of April to see the aircraft first-hand. Inspection confirmed they were in good condition, and so airframes 352 & 355 were purchased, along with two spare low-hour Viper engines.

In the spring of 1983 Mike and Eric surveyed a number of surplus Singapore Air Force Jet Provosts. 355 was one of those selected (later becoming G-JETP), along with 352 (G-PROV). Both were dismantled and returned to England by sea. (EGH)

Although not a pilot, Eric often found himself in the cockpit of one of the Hunters, undertaking ground engine runs and sometimes moving an aircraft around the apron. Here he checks the positioning of G-HUNT on the flight line at the 1982 Bournemouth Air Show. (EGH collection)

> *Eric:* Flying by Singapore 747 to Changi, I accompanied Mike on the trip since he needed an engineer to check over the Jet Provosts. I considered that they needed little work to restore them to flying condition, and so Mike went ahead with the purchase. I then had to make arrangements for the aircraft to be crated and shipped back to Bournemouth later in the year. During the flight Mike talked about setting up a dedicated organisation to look after his increasing jet fleet. He asked if he set up an operating company, would I be its Chief Engineer to look after the aircraft. Although I had a job with Airwork, Mike's plan appealed to me and so I said "Yes". Hooked once more! So, I left Airwork at the end of May and started working for Brencham Historic Aircraft on 1 June 1983.

Mike, as Chairman of Brencham, said it was no problem to form a new company within the Group – setting up Brencham Historic Aircraft Co. The new company's aim was to preserve British jet fighters in flying condition and display them at air shows. Although Stephan had originally flown much of the display work, he did not join the new set-up. So, as well as his day job as a 737 captain with Britannia, Adrian became team manager and senior display pilot. Restoration work would take place at the Bournemouth engineering and operational base, with the airworthy fighters housed in a new complex built for Brencham in 1983 at Biggin Hill. This comprised the administration headquarters 'Patterson House', plus the executive aircraft division and hangar for the historic aircraft. In Patterson House's board room there was a painting of an imposing looking gentleman hanging on the wall. This was referred to as Lord Richard Brencham, the Scottish founder of the group. In fact there was no such person, but Mike felt it added something to the image of the business!

With an operating company formed, two airworthy Hunters owned and other jets in the pipeline, it was time to think of the 1983 season. A catchy title was needed. The display work had been the sole preserve of G-HUNT, which used the radio call sign 'Hunter One'. This seemed a suitable title for solo operations, but these would now include G-BOOM. In spite of there being two aircraft, the name Hunter One was agreed on.

4

Hunter One

Hunter One was now in business, with the aim of displaying to the public its two (at the time) unique Hunters. There was strong interest from air display organisers during 1983, with bookings handled through Brencham's London office. Early in the season was a warbirds display at Stapleford Tawney (flown from Stansted) on 1-2 May, then there was the Biggin Hill Air Fair on 14-15 May (with G-HUNT also flying up to Barton's Show), with G-HUNT and G-BOOM both calling in at Luton on 29 May. As well as the UK, there was interest from the continent, resulting in two displays during June. The first involved G-HUNT being displayed at Lille by Adrian on 5 June, with Mike flying it to and from the show. Then a display at Bordeaux on 22 June, again requiring the fitting of G-HUNT's four underwing fuel tanks for the transit flight south. The display at BAe Dunsfold by Adrian on 18 June resulted in a letter of thanks to Mike from John Farley, saying they had all been pleased to see G- HUNT again. Other displays later in the summer included Brands Hatch, Cowes, Maastricht and Utrecht.

> *Adrian:* Mike had flown HUNT to Bordeaux for their show but was hit by lightning returning home, requiring a diversion into Evreux. I had a call from Mike to say that he was not happy with the feel of the Hunter, and would I fly out to have a look. I arrived on 24 June, Mike having already made his way home. Ground running seemed satisfactory, but I had to reject my take-off when I spotted the airspeed indicator was not working. But as I slowed down to turn off the runway it started to give a reading. So I set off again and, although there were fluctuations, the indicator was partially working. I pressed on and made it safely back to Bournemouth. On examination Eric found that part of the inside of the pitot tube had melted following the lightning strike, hence the reason for a false reading as I sped up for take-off. Once airborne it worked OK.

For the increased display flying, additional pilots were needed by the Hunter One Team, with Geoff Roberts joining for the 1983 season. Geoff had been a QFI on Jet Provosts, TWU Instructor at Brawdy on Hunters and Hawks, 1983 Tornado display pilot, as well as being a current Tornado instructor – more than enough

Normally G-HUNT flew with two under-wing fuel tanks to reach UK air shows. The addition of four tanks easily gave it the range to reach air shows in further parts of Europe such as Austria and Switzerland. (EGH)

The Hunter One team of 1983, posed in front of flagship Hunter G-HUNT. Eric (Chief Engineer), Geoff Roberts (Display Pilot), Mike (Owner), Adrian (Team Manager/Senior Pilot) and Brian Henwood (Display Pilot). (Peter R March)

qualifications to join Hunter One! Geoff was joined a few months later by Brian Henwood, who was a CFS examiner, QFI, RAF Phantom display pilot 1980-81, and also currently flying Tornados. Again, another eminently suited candidate. G-BOOM displayed at Woodford on 25 June and Humberside on 3 July. The Bournemouth Air Pageant of 17-18 July enabled Hunter One to display both G-HUNT and G-BOOM to the Bournemouth locals. Also flying was Spencer Flack's former Spitfire, G-FIRE. G-HUNT and G-BOOM were then displayed at RAF Valley on 13 August.

> *Eric:* For the Humberside Show in July Adrian flew north, with me as his 'passenger'. In all the years that I had worked at Dunsfold, this was the first occasion that I had flown in a Hunter – if only for thirty-two minutes! However, it still failed to change my view that I'd sooner be an engineer than a pilot!!

As mentioned, the name always associated with the Hawker Hunter is that of its famous test pilot, Neville Duke. Neville had retired to the Lymington area and flew his own Cessna from Bournemouth. Mike Carlton saw an opportunity for some publicity for Hunter One in connection with the Thirtieth Anniversary of Neville's World Air Speed Record of 7 September 1953. The record had been conquered in a red Hunter, and here was Hunter One flying red Hunters. Mike contacted Neville, now 61, and offered him a flight in G-BOOM to commemorate the record date. So on 7 September Neville and his wife Gwen arrived at Hangar 600, along with plenty of media personnel. Mike flew Neville over Tangmere and the Sussex coastal route again at 300ft to see how close to the record speed of 727mph they could reach. Mike was overjoyed to attain 700mph, thereby raising £700 from sponsorship towards the Stoke Mandeville Hospital Appeal. For the flight G-BOOM was accompanied by G-HUNT flown by Adrian. It was Neville's first Hunter flight for some years and he commented, 'It went like a dream, and I think we surprised one or two fishermen off the Sussex Coast!'

Further publicity for Hunter One followed later in September, with a London to Paris Air Race. This was similar to a 1959 race, the aim being to attain the fastest city centre to city centre time. The race was preceded by a press and TV presentation at Cranfield on 22 September, and G-BOOM was used by one of the teams for their record bid on 24 September. The competition included teams using an RAF Gnat and Hunter Trainer, as well as a civilian Citation executive jet. Operating from Biggin Hill, Mike was to fly Tim Ridgeway as team 'runner'. Arriving by A.109 helicopter, Tim jumped into the already running G-BOOM, although Mike was not allowed to move until Tim was strapped in. G-BOOM's morning attempt failed to break a 1959 record due to French air traffic problems delaying their arrival at Le Bourget. However, the evening return run with David Boyce as 'runner' set a new record of 38 minutes 58 seconds between the Arc de Triumph and Marble Arch. G-BOOM flew the Le Bourget to Biggin Hill sector at low level, resulting in its paintwork rippling! (it was repainted the

Neville Duke and his wife Gwen share a glass of wine after the September 1983 anniversary flight in G-BOOM. Piloted by Mike Carlton, the Hunter almost attained the same speed as Neville had achieved thirty years earlier. (British Aerospace)

following February). Gatwick Radar greatly assisted their return, the weather was atrocious, bringing G-BOOM straight in line with Biggin's runway, where Eric was awaiting its arrival. Helicopters and motor bikes provided onward connections for Mike and David to the capital. They leapt into a waiting Army Lynx helicopter, leaving the Hunter on the perimeter track with engine still running and tail parachute dangling from the rear of the aircraft. As he slammed the helicopter door shut, Mike's parting words to Eric were, 'park it for us mate'! On arrival in London, Mike and David switched to motor bikes for the final leg to Marble Arch. Hunter One obtained publicity as the event was shown live on Noel Edmonds Saturday evening 'Late Late Breakfast Show', the team having flown to the TV centre by another Lynx. The press picked up on G-BOOM being the world's fastest private two-seat 'executive jet'. The record-breaking event was celebrated by a reception laid on by Mike at the 'Inn on the Park' in London on 12 October; it later entered the Guinness Book of Records. The 1959 event had been won by an RAF Squadron Leader also flying a Hunter.

As if he did not have enough on his plate, Mike Carlton was on the look-out for even more aircraft to expand his fleet. In May 1983 the collection of the Historic Aircraft Museum at Southend was put up for auction by Phillips. Mike attended and ended up purchasing Meteor T.7 VZ638 and Sea Hawk 'XE364'/XE489. The auction catalogue indicated that both were in reasonable condition and suitable as restoration projects. Initially they were moved by road to Eastleigh for storage in the hangar of Norman Bailey Helicopters – another company that Mike had just bought. Consideration had been given to using Eastleigh as a base for some of the Brencham Historic Aircraft work, but Mike

The London to Paris air race team at Biggin Hill in September 1983. Tim Ridgeway (runner) and Mike (pilot) in the cockpit of G-BOOM, with Gary Savage (helicopter pilot) and Eddie Kidd (motorcycle) in front of the Hunter. (A Gjertsen collection)

Meteor T.7 VZ638 was purchased by Mike at the Southend museum auction in May 1983, the airframe being described as within the realms of restoration. It arrived at Bournemouth in November 1983, initially receiving a cosmetic tidy-up. (MHP)

At the Southend auction Mike purchased a Sea Hawk. He also spotted a second Sea Hawk under restoration, and concluded the purchase of WM994 in the autumn of 1983. When problems were discovered with the first aircraft, restoration work was concentrated on WM994. (MHP)

The expanding Hunter One fleet meant more work for the engineers. By 1983 Eric had been joined by Ivan Henderson and David Greaves, with Bob Tarrant seen here strapping Adrian into one of the Jet Provosts. (Peter R March)

bought Glos-Air in October 1983, so retaining the work at Bournemouth. However, it was now known as Glos Air Aviation, without the hyphen. The Meteor and Sea Hawk moved to Bournemouth by road in November. A second Sea Hawk G-SEAH (WM994), which had been in storage at Southend, was purchased and moved to Bournemouth in October. Further swelling the ranks, the two Singapore Jet Provosts arrived at Southampton Docks in containers on 1 November, and were moved to Bournemouth a few days later. In December they were allocated the appropriate civil registrations G-JETP and G-PROV. At the time it was not realised that most of the new fleet had Bournemouth connections. Sea Hawk 'XE36' had been operated from Bournemouth as XE489 in the 1960s by Airwork's Fleet Requirement Unit. In the summer of 1969 both the Jet Provosts had been delivered through Bournemouth to their original operator, the South Yemen Air Force.

From the autumn of 1983 operations by Hunter One at Bournemouth had competition from across the main runway. FR Aviation maintained six F-100F Super Sabres on behalf of Flight Systems of Mojave. They had a contract to operate them for the USAF Europe for aerial combat practice with their European based fighters. The F-100's take-offs were noisy and there were some interesting landings over the years.

During the winter of 1983/84 G-HUNT was fitted with wing leading-edge extensions, as fitted to the Hunter F.6, and a tail parachute. It was test flown at the end of March. As built, it had a straight-wing leading edge, whereas the F.6s delivered to the RAF were fitted with a sawn-tooth leading edge for better handling. G-HUNT was further modified in the autumn of 1985 to produce white smoke for the 1986 display season, the necessary thirty gallons of diesel fuel being carried in the forward portion of one of its modified drop tanks. This was then piped through the wing and fuselage to the jet exhaust pipe. This system differed from that used by 111 Squadron thirty years earlier, as they carried the diesel in the Hunter's modified gun pack. Brencham developed a new Avon engine starter system, based on one used by Rolls-Royce on its static-engine test rig. The original cartridge starter system was replaced by a new electrical system powered from additional batteries housed in the fuselage. Initially a primitive electrical starter was fitted to G-HUNT on its rebuild at Elstree, but now a far more efficient transistorised and fully computerised system was fitted. This was more suitable, and became standard on most of the Hunters subsequently refurbished at Bournemouth. Up to now Eric and his team of engineers had just G-HUNT and G-BOOM to keep flying. They were now faced with major engineering work on a number of airframes.

Citation G-UESS of Mike's TransEuropean Air Charter company crashed into the sea on approach to Stornoway on 8 December 1983. This was the result of a misjudged night-time approach, and all on board were lost. The dead included Tim Ridgeway, who was one of the runners on the London to Paris Air Race. Shortly afterwards TransEuropean was closed down.

Mike pursued his aim of having a fleet of airworthy British jet aircraft by making further acquisitions during 1984. He sought examples of the de Havilland Vampire, first buying, in March, T.11 XK623, which was in storage at Moston Technical College in Manchester. As part of the deal, the college wanted a replacement aircraft. Eric knew of a Tri-Pacer which had been blown over in a gale at Bournemouth the previous summer. So the wreck was purchased, tidied up and despatched at the end of April. It did not survive long as it was destroyed by fire after eighteen months. A second Vampire XD599 was found in storage with Doug Arnold's collection at Blackbushe and was purchased, arriving at Bournemouth by road on 15 March. It was followed by XK623. However, the main part of the deal with Doug Arnold was the purchase of Gloster Meteor WM167, more of which later.

Restoration work on the two Jet Provosts commenced during the spring of 1984, when the logbook of G-PROV revealed that it had suffered minor bullet damage. This was from rifle bullets whilst the aircraft in service with the South Yemen Air Force in September 1969, so proving to be a real warbird. There were holes of a different kind affecting Sea Hawk 'XE364' (now registered G-JETH). On dismantling the aircraft, it was discovered that the main wing spar connections to the fuselage had all been drilled through. This had been in the terms of its original Royal Navy sale contract to the Southend Museum back in 1967, something that Mike was unaware of when he made the purchase. The auction catalogue had made no mention of the deficiency, whereas a Mitchell in the auction was clearly described as having its main spars cut. (The author had been aware that the Sea Hawk had been rendered un-airworthy, thinking

The spring of 1984 saw the purchase of two Vampire Trainers for the Hunter One collection. XK623 '56' came from a college in Manchester and XD599 'A' from Doug Arnold at Blackbushe. Lack of hangar space meant they spent most of their time in external storage. (MHP)

that Brencham would also be aware of the problem). WM994, the second Sea Hawk was re-assembled and painted in primer. The two Vampires were also re-assembled, with XE364 gaining an unofficial personalised registration 'G-VAMP', which already belonged to a 'Thunder' hot-air balloon.

The 1984 display season included two eventful trips to the continent. Both G-HUNT and G-BOOM flew to Liège for their air tattoo on 2-3 June. During the second display G-HUNT hit a seagull, which resulted in Adrian landing with a big dent in the nose cone. This necessitated an SOS phone call to Eric's home for help to enable G-HUNT to get back to England.

> *Eric:* Late on the Sunday afternoon I received a call from Adrian asking for my help following the incident. So I drove down to Bournemouth Airport to find a spare nose cone in our stores. Then there was the problem over its size, which was too big to fit in a Cessna or Cherokee. So Mike chartered an Agusta 109 from Alan Mann Helicopters and I managed to get the nose cone on board. Unfortunately, the Agusta could only take me to Brussels, the pilot having to return immediately for another charter. So there I was airside at Brussels, not speaking the language, having no customs papers, no currency, only my passport, a toolbox and a nose cone. With horror I thought to myself "Help, what am I doing here?". I spent some time persuading an elderly Customs Officer that I had not stolen the aircraft part (there were no import papers), and that I had just flown in from England – "But where is

The first of the former Singaporean Jet Provosts to receive attention was G-PROV, seen in the Glos Air hangar during the summer of 1984, with G-JETP behind. Note the murals of Aero Commanders painted on the wall – Glos Air being the UK Agent at the time. (Peter R March)

your aircraft??" "It's already flown back to England" I replied. After half an hour he let me through, muttering, "the English are mad" as he walked away.

I went to the train station and purchased a ticket to Liège, but found I could not get the nose cone through the carriage door. Back to the bus station, where first I tried to get the nose cone onto a bus. No luck, so off to the taxi rank. But it would not fit in the normal taxis. Luckily I found a large Mercedes where I could strap it in the boot, although the lid remained opened. Off to Liège Airbase by which time it was getting dark and everyone had disappeared to their hotels. Luckily the Hunter was in a hangar, so I went to work and replaced the damaged nose cone. Adrian had left a note "Have gone to my hotel – will see you there". I found someone to drive me to the hotel where I met Adrian and told him the Hunter was fixed and would be able to fly home the next day. Then I went to bed!

I abandoned the dented nose cone in Liège, returned to Brussels, where I flew with my trusted toolbox by One-Eleven back to Heathrow and home. As ever, Adrian had flown wonderful displays both days, and returned to Bournemouth before I did! – Pilots!

The following weekend G-HUNT and G-BOOM flew south on 11 June for the Caen D-Day Commemorative Air Show, with the support team using an Agusta A.109. There was much French press coverage of the arrival of the two Hunters, which at the time were still a rare combination. However, there were problems for G-BOOM on landing.

Eric: I had arrived the day before by A.109 and so was mingling with the press for the aircraft's arrival. They did their usual impressive fly-past but, on landing, I noted that BOOM was right wing low as it taxied along the runway. As it turned off the runway it was evident that there was a problem. HUNT arrived on the apron first and I ensured that BOOM was then marshalled behind it. As the crews disembarked I shouted to them "Go to HUNT and keep the press with you" "Why" they retorted – typical pilots! "Don't ask questions – do it" I shouted back.

The pilots gathered around HUNT and posed for lots of photos and interviews. The press were happy and after they had gone away I was able to show the pilots the problem. "Have a look at BOOM's starboard undercarriage leg". The pilot of BOOM went visibly white. During its landing the casting on the leg had split open vertically and the oil had drained away, but luckily the leg had not fully failed or else we would have been out on the runway recovering a badly damaged Hunter. We flew an extended display with HUNT only the following day, leaving BOOM in the static line up with a thick piece of wood propping the undercarriage leg up! Luckily no one asked why.

I flew back to Bournemouth by the A.109 and collected a replacement leg and two engineers. All three of us returned to Caen as soon as possible to rectify the problem. After delays in finding a suitable jack, we changed the leg the following day, refilled the hydraulics and bled out the system. On the Tuesday BOOM flew

G-HUNT and G-BOOM seen on the eventful trip to Caen in June 1984. Only HUNT took part in the flying display, with G-BOOM parked up in front of the terminal. Close examination shows it as starboard-wing low due to the cracked undercarriage leg. (Brian S Strickland)

back to Bournemouth with an engineer as passenger – the other two of us returned by Navajo. The press coverage of the display was comprehensive and appreciative. Fortunately no one knew that one of our Hunters had broken a leg, and our small deception still remains undiscovered to this day.

Back home, Mike added a Swift and Venom/Sea Venom to his shopping list, resulting in a visit to Duxford in June to inspect some former Swiss Air Force Venoms. However, they were not considered suitable. Doug Arnold finally agreed to sell his Gloster (AW) Meteor WM167 which had been in storage at Blackbushe for some years. Little work was found to be needed on the Meteor to make it airworthy for a ferry flight down to Bournemouth for a full overhaul.

Eric: In June the task of making WM167 airworthy commenced using some Glos Air personnel. All proceeded smoothly, and an application was made to the CAA for a permit for the ferry flight. Engine runs were demonstrated for their surveyor, who issued the permit in July. Brian Henwood, a former Meteor pilot and one of our new display pilots, was flown by our A.109 to Blackbushe on 6 July with instructions to undertake the necessary engine runs and taxi trials. Mike gave him the OK to undertake a short air-test, if he felt happy with these. During this activity a watching brief was kept from the hovering helicopter, which maintained radio contact with Brian. However, he took us by surprise, for having started both engines, he taxied to the end of the runway, called the tower for take-off clearance, opened the throttles and was away. But not on a short air-test – he was off to Bournemouth! We chased him as fast as we could, and on arrival at the hangars Brian explained that, "it all just felt right" and so went for it, even offering to take the Meteor up again for a demonstration.

By the summer of 1984 Meteor TT.20 WM167 had been restored to its NF.11 configuration, flying again in the markings of 141 Squadron. It soon became a firm favourite on the air display circuit in the UK and the Continent. (Andrew March)

When Doug Arnold had acquired the Meteor it had been operated as a TT.20 target-towing version. Most of the equipment was removed to return the aircraft to its original NF.11 night-fighter variant, and in reality it only needed minor attention to make it fully airworthy. However, it was given a major servicing before joining the two Hunters on the display circuit for the summer of 1984. It was repainted to represent a NF.11 of 141 Squadron, also carrying its civil registration G-LOSM (GLOster Meteor). This aircraft also had local connections, as it had been a trials aircraft with Flight Refuelling at nearby Tarrant Rushton for many years. As well as the new aircraft, a large quantity of major spares was acquired from various locations, including Derwent engines from Twente and Nene engines from Bergerac. Eric and his engineering team had more than enough work to keep them busy, needing at least an eight-day week to get it done!

Bournemouth's 1984 Air Show was greatly expanded, as it was now organised by the International Air Tattoo team of Greenham Common fame. Backed by local television company TVS, and with greater publicity, there were many more aircraft on display. The show's souvenir programme was edited by Peter March, who called on Eric for background information on an article being included in the programme. Already acquainted with Hunter One operations, Peter became a regular visitor to Hangar 600 to record events. The Hunter One Collection was displayed at the show on 18-19 August so that the public could see the restoration work already undertaken. G-LOSM joined G-HUNT and G-BOOM in the flying display, having received its CAA approval the previous week, with the other aircraft on static display under the Brencham Historic Aircraft banner. Here they were surrounded by the new generation of fighters – Harriers and F-16s.

The 1984 TVS Air Show South at Bournemouth enabled aviation artist Wilf Hardy to produce a programme cover which featured G-HUNT and G-BOOM in formation with a RAF Harrier GR.3 over the main runway. (RAF Benevolent Fund)

At the TVS Air Show, Hunter One were able to display its collection to the local public, with most of the aircraft currently under restoration shown in the static park. At the time G-PROV was still three months away from its first flight. (MHP)

Vampire T.11 XK623 was also on static display at the 1984 show. By this time its RAF serial had been removed from the boom and replaced by civil registration G-VAMP. This was the Hunter One engineers seeking a personalised registration for the Vampire. (MHP)

Adrian: As Team Manager I devised the various training and display routines, and also made the necessary arrangements with show organisers. For an example of the routines, the full eight-minute display for G-HUNT was as follows:

Run in at 500kts for curving turn with throttle at idle
Pull up 6g, reversing high and turning back for a barrel roll
Pull up for a barrel roll at 320kts achieving 4½g in the vertical
Wingover to run in parallel to crowd line for slow roll at 300kts
Full power, pull up for half Cuban, using 5g and topping at 3,500ft, decelerating turn to pass along crowd line with flaps and undercarriage lowered at 150kts
Gear retracted during 360° flat turn, pulling round to fly-past inverted at 300kts
Reversing high for a barrel roll at 320kts, accelerating in wide turn for a final low pass in excess of 500kts.

Show planning frequently meant dealing with overseas organisers. For example, I flew G-HUNT to Klagenfurt, Austria, on 31 August for a large show there on 1-2 September. This meant getting the necessary authorisation to overfly Belgium and Germany, as well as the final leg within Austria. Then the display had to be agreed with the organisers. I also anticipated that the return flight would include a refuelling stop at Liège. However, I remember Klagenfurt for a different reason. The show included many high-profile displays of which I was one. All pilots were well looked after by the organisers and I noticed a number of them around the Pilot's Tent after I landed on the second day. Asking what they were waiting for, I was informed, "our payment". Word had got around that the show had not raised enough money to pay the bills, but some cash was available to those pilots who turned up first! Luckily, from my point of view, there were still enough Schillings on the table when I reached it, although they ran out a few pilots later. I felt almost embarrassed as I walked past fellow pilots, stuffing the notes into my flying suit.

Eric: There were regular visits to our hangar by an aircraft engineer who used to work on G-LOSM when it was operated by Flight Refuelling. He seemed surprised when I mentioned the long trip to Klagenfurt. "How did you cope with the distance?" he asked. I explained that we fitted long-range tanks for the extra distance. "Didn't that hinder the display routine?", he queried. So I told him that our engineer flew down in the back seat, and prior to the display he removed the tanks. The engineer still expressed surprise that we got back without problems. On being told that G-LOSM cruised home at 30,000ft he exclaimed horror, saying that Flight Refuelling only ever let it reach 10,000ft!

In June 1984 Brencham purchased the Bournemouth-based Shirlstar and Via Nova companies, mainly for their large hangar on the north side of the

The Hunter One fleet were occasionally used for film and advertising purposes. Two fighters were required in 1984 for the making of *Return to Fairborough*, and Hunter One were able to provide their two Hunters. Pseudo-military markings were applied, as seen on G-HUNT. (MHP)

The return of G-PROV to the air at the end of 1984 was a suitable occasion to obtain some publicity photographs of the Hunter One fleet. Seen high over South Dorset are G-LOSM, G-HUNT and G-PROV. (Peter R March)

airport. In due course it also housed Brencham's main spares store. The group expanded its activities further with the acquisition on 31 August of the majority shareholding of Bournemouth-based Metropolitan Airways. The airline operated scheduled services from Bournemouth northwards to Newcastle with a fleet of Twin Otters and Short 330s. To keep the business in house, these now visited Hangar 600 for maintenance.

During the summer a request was received from a film company which required aircraft for a TV film they were making, staring Robert Mitchum and Deborah Kerr. The story involved a former USAAF pilot returning to England to relive his wartime experiences, ending up with his making an unauthorised flight in a T-6 Harvard. Two fighters were needed to force him to land, with G-HUNT and G-BOOM proving ideal for the requirement. However, CAA regulations meant they could not fly in RAF colours, as requested by the film company. So instead of red, white & blue roundels, they were painted white, red & blue. Filming of *Return to Fairborough* commenced at Fairoaks in the middle of October, with the two Hunters flown by Adrian and Geoff, chasing Harvard FT323 supplied by Gary Numan, all filmed from an Agusta A.109. After one day's filming the Hunters flew a low pass along Fairoaks' runway as the Harvard landed. In fact it was so low that they almost forced the Harvard into a ground loop! Further filming took place from Bournemouth during November, using a Heron as camera ship (the author recalls being forced to leave the Sunday lunch table quickly as the Heron and Harvard flew over, soon chased by the two Hunters).

The first of the Jet Provosts G-PROV was completed by November, enabling Adrian to make a test hop on 23 November, followed by full flight on 27 November. At the time it was in primer, but was soon painted red to match the Hunters. This enabled it to take part in a four-ship Arthur Gibson photo shoot on 14 December. Luckily the weather was perfect, with the Hunter One fleet being photographed from Arthur's Aztec G-FOTO whilst flying over the Purbeck Hills.

Adrian: The arrival of the Jet Provosts gave rise to plans for Hunter One to introduce civilian pilots on to military training jets, before progressing to the Hunter. I drew up a training plan for consideration by the CAA but it was not accepted. The problem was that the Hunter One aircraft flew on Permits to Fly, not C of As, and as such were unable to be operated for hire or reward. Ron Ashford, Director General of the CAA Safety Regulation Group, understood what we wanted and tried to help. He was instrumental in enabling HUNT to fly IFR instead of VFR, something that was normally only granted to an aircraft with a C of A.

5

Increased Display Flying

After his commemorative flight Mike maintained contact with Neville, resulting in the following appearing as a foreword to the Hunter One publicity brochure:

> It is fortunate that there are far-sighted individuals with the initiative and resources to collect and preserve flying examples of the best British aircraft, before they become extinct. The Hunter One Collection is unique in being all jet, consisting of fighter/trainer aircraft spanning 40 years of aviation history. The parentage of the Meteor Mk.7, oldest type in the Collection, dates back to the 1943 Meteor Mk 1, fitted with Whittle-designed Welland engines. Meteors entered RAF service with 616 Squadron the following year and saw combat against the "flying bombs": the last survivor in RAF service was retired in 1983.
>
> The Meteor established the first jet-powered World Speed Record at 606 mph in 1945 and increased it to 616mph the following year. The Meteor's successor in RAF service was the Hawker Hunter Mk.51, founder-member of the Collection and a popular favourite at displays. It has been joined by a Mk.53 two-seat trainer. Although these are "civilianised" Hunters – the first in the UK – the type is still very much in operational service with a number of overseas air forces. Nearly 2,000 Hunters were built, many being subsequently "recycled" after overhaul and conversion.
>
> Having flown the first Hunter on its maiden flight in 1951, it was an especial privilege and pleasure to fly one again 32 years later. It also enabled me to appreciate at first hand the meticulous standards of maintenance and professionalism which make the Hunter One Team such excellent guardians of an important part of our aviation heritage.
>
> Neville Duke.

In addition to the main Hangar 600, Hunter One also had use of an adjacent blister hangar, which had been left over from the Second World War. This was used for storage of parts, including the stripped out fuselage of Hunter ET-272

Early in 1985 a spare Hunter airframe was prepared at Bournemouth for gate guard duties at Brencham's HQ at Biggin Hill. After completion, but before painting, the engineers used an aerosol can to apply the unofficial registration G-ERIC. (MHP)

After painting, the gate guard Hunter was moved by lorry to Biggin Hill and placed on a plinth outside the entrance to Patterson House. The base of the fin carries the letters BHAC – Brencham Historic Aircraft Company. (JHL collection)

Adrian happily accepting some pre-flight advice from Eric before a Jet Provost test-flight. A close working relationship had built up between the two warbird enthusiasts. Note that Adrian's flying suit carries a Hunter badge with the wording 'Last of the Sports Models'. (Peter R March)

– less its cockpit and nose section. Redevelopment meant that Bournemouth Airport wanted the blister hangar removed, whereupon Mike developed an idea for the Hunter. It could act as a gate guard to the new Brencham HQ at Biggin Hill. So, the fuselage was brought into the main hangar, a fighter nose was added, plus spare wings and a tailplane. Prior to painting red, the airframe was painted with the fictitious registration G-ERIC by the engineers, but the finished aircraft carried the marks BHAC (Brencham Historic Aircraft Co.) in small letters on its fin. The various sections departed by road for Biggin Hill at the end of March, being re-assembled in April on a plinth outside the HQ reception entrance.

For 1985 Adrian continued as Team Manager/Chief Pilot, the Hunter pilot was Geoff Roberts and the Meteor pilot Brian Henwood. Mike's original reason for buying G-BOOM was so that he could convert to being qualified Hunter pilot. Having achieved this aim, he was able to fly G-HUNT at displays, with G-BOOM now considered surplus to requirements. So, in the spring it was prepared for sale to the United States as a high-speed photographic aircraft, but the deal fell through at the last moment. The search for a Swift was having results, with XF114 found in use as an instructional airframe with the College of Technology at Connah's Quay, Flintshire. The college no longer needed the old fighter, saying that if they disposed of it then a suitable replacement airframe would be required. So, non-airworthy Jet Provost XR654 was acquired by Brencham at the end of 1984 for possible exchange, whilst negotiations for the Swift continued with the college during 1985.

Air show appearances included the Biggin Hill Air Fair on 11-12 May. For this G-HUNT and Jet Provost G-PROV were despatched, and they flew in formation with Spencer's former Spitfire G-FIRE – a scarlet trio. Unfortunately, the weather during the show was wet with low cloud. G-HUNT and G-BOOM displayed at Mildenhall's Air Fete on 25-26 May, which resulted in another incident with G-HUNT. Problems with the port wheel brake on the first day of the show resulted in a burst tyre, with the Hunter veering off the runway towards the end of its landing run. This was another occasion when Eric had to sort out the pilot's problem, taking an engineer with him and working through the night. The tyre and brakes were replaced to enable G-HUNT to fly out the next day.

The summer of 1985 proved to be busy, with G-HUNT displaying at the Fighter Meet at North Weald on 29-30 June, flying all the way to Vienna on 6 July and displaying at Alconbury's Open Day on 21 July. 25 July saw G-HUNT and G-PROV appear at Brawdy's Open Day. G-LOSM and G-PROV flew to the Ursel Air Show in Belgium on 3-4 August. However, the Meteor ran into problems at Ostend Airport on its way home on the Monday.

Eric: I was gardening at home, and at about 5.00 p.m. I received a phone call from Brian Henwood, who explained that he and Mike had stopped at Ostend to refuel. However, as Brian put it, "We're outside, can't open the canopy and can't get into the bloody cockpit! What do we do?" I made several suggestions, amongst which was "Start walking". Whatever they tried had no success, the outcome being that I had to get to Ostend to

At the Biggin Hill Air Fair in May 1985 G-HUNT displayed alongside its former Spence Flack stablemate Spitfire G-FIRE, as well as Jet Provost G-PROV. Spencer would have appreciated the formation – a scarlet trio. (Peter R March)

> rescue them. After several phone calls I found a pilot who would fly me and my toolbox in his Cherokee from Bournemouth. It took 1 hour 10 minutes to fly there and it was 8.30 p.m. when I eventually arrived.
>
> Sure enough, the canopy was firmly locked closed. Impossible to do, I thought! Inspection by torch showed that Mike, who was in the rear seat, had fiddled with the canopy release handle and left it dangling down in the locked position. Having located the problem, Mike jumped into the Cherokee and was on his way back home. I never even saw him go!
>
> I borrowed a drill from the local workshop and, with a length of welding rod, was able to eventually unlatch the canopy. By the time I had returned the tools, Brian was in the cockpit with engines running. As well as failing light, it had now started to rain. So I climbed aboard to find the rear cockpit full of bags and equipment. By this time we were taxiing out and were airborne before I had time to settle down amongst the bags and strap myself in. We had a safe, if somewhat late, return to Bournemouth that evening.
>
> Two things came of the incident: (i) I never got any gardening done; (ii) A large label was attached to the canopy release handle, "Do not touch, and never let your dingle dangle".

A request was received from airshow organisers in Linz in Upper Austria for one of the Hunter One aircraft to appear at their display on 10 August. Despite its problems on the previous weekend, G-LOSM departed on its longest ever flight on 9 August for the display. This required the fitting of the Meteor's belly fuel tank as well as the two underwing drop tanks. Luckily, there were no problems this time, the crew finding themselves in the company of two RAF Harriers and two Jaguars.

On the restoration front it was hoped to have Sea Hawks G-JETH and G-SEAH flying by the end of 1985, with Meteor VZ638/G-JETM following in 1986. A solution still had to be found to the problem of G-JETH's cut spar attachments. The Sea Cadets at Chilton Cantelo, just south of Yeovilton, had Sea

Problems with restoration of the two Hunter One Sea Hawks meant that a third airframe arrived to provide spare parts. This summer 1985 view shows G-JETH, WM983 and G-SEAH lined up outside the Glos Air hangar at Bournemouth. (MHP)

To boost its fleet of naval fighters, Hunter One only had to look across the runway at Bournemouth where Sea Vixen XS587 had been used for trials by Flight Refuelling. Registered G-VIXN, hardly any work was undertaken before the restoration plans were ditched (MHP)

Hunter One's second Jet Provost was first flown by Adrian in July 1985. In contrast to other aircraft in the fleet, G-JETP was later painted in an indigo blue scheme in order to undertake display work, usually in company with G-PROV. (Andrew March)

Hawk WM983 on display at their HQ, and it was arranged to 'borrow' their aircraft to help XE489/G-JETH. The plan involved changing the wings over and then switching identity of the aircraft. So, the wings of XE489 were attached to the fuselage of WM983, which then assumed the identity of G-JETH. The fuselage of XE489 (with the cut spar) was repainted as WM983 and returned to the Sea Cadets in October. Their only requirement was that the aircraft's hydraulics still worked so that the fighter's wings would fold up and down.

Meteor VZ638 was cosmetically restored and repainted in standard RAF colours, but still needed a lot of work to make it airworthy. Meanwhile it was agreed to relegate the two Vampires to longer-term projects. The second former Singapore Jet Provost G-JETP first flew in July 1985, later being painted indigo blue, with gold letters and trim. Mike still had ideas for expanding the fleet, now considering a Lightning and Sea Vixen. Still being in RAF service at the time, no suitable Lightnings were available, but there were some surplus Sea Vixens being used for trials work by Flight Refuelling just across the runway at Bournemouth. A number were up for disposal, and a dealer offered one to Mike. A deal was struck, and in August XS587 moved the short distance to the Glos Air hangar to be appropriately registered as G-VIXN. Mike was keen to have the Vixen airworthy as soon as possible, certainly in time for the 1986 Air Show season. To speed up work Mike arranged for some Flight Refuelling engineers to come and assist Eric if necessary.

By now Glos Air Aviation had moved into the biz-jet overhaul business, with BAe 125s and Citations arriving for attention. To house the increasing workload a large new hangar was built adjacent to Hangar 600, shared by Brencham and Metropolitan Airways, being completed during October 1985.

6

'... *And Then I Get A Pilot*'

Hanging on Eric Hayward's office wall in Hangar 600 was a plaque with the quote '... and then I get a Pilot'. The story behind it relates to the different conceptions of flying vintage jets – that of the engineer and that of the pilot. Eric had spent many years working as an engineer for Hawker's at Dunsfold, tasked with preparing Hunters, Harriers and Hawks off the production line for flight. The pilots would arrive for 'work', jump in the aircraft and take off for a test flight or display routine.

> *Eric:* You will have noticed my comments about pilots – this helps to explain why. During my early days as Chief Engineer at Hunter One I was often interviewed for radio and television. Asked to describe what was done during the restoration of an aircraft, I replied that when a new aircraft arrived it was usually in a poor state of repair. So, after dismantling and steam cleaning, I undertook an inspection to determine what repairs and replacement parts were required. In due course the aircraft would be re-assembled and tested, the fuel system checked and engine fitted. Engine runs and taxi-tests would follow, then when all was absolutely and entirely to my satisfaction, the aircraft would be prepared for a test flight. Then of course what did I say – "and then I get a Pilot".
>
> The Hunter One pilots, particularly Geoff Roberts, took umbrage at my phrase, pointing out that he was not just 'any old pilot' but a technically skilled professional. My stock response was that pilots were only a part of the restoration exercise – like filling the hydraulics with oil or putting air in the tyres – the aircraft could not be completed without these and many other operations. Pilots always need to be brought down to size!
>
> However, at a celebration party in August 1984 the Hunter One pilots presented me with an inscribed wooden plaque which I still cherish at home to this day. The two types of minds were beginning to think as one!

7

Retrenchment

August 1985 brought a downturn in Hunter One's dynamic operations, as Mike Carlton realised that operating vintage jets required a bottomless pit of money, which even he did not have. His various property businesses within the Brencham Group were also suffering financial problems at the time. Initially Mike had raised capital by mortgaging the aircraft, but on 22 August he announced a cut-back in Brencham Historic Aircraft operations and the disposal of some of the fleet. He decided to retain G-HUNT, G-LOSM, G-JETP and G-PROV, to continue with the restoration of G-SEAH but suspend work on G-VIXN. The rest of the fleet were then advertised for sale. Within any business financial arrangements are often complex. Brencham was no different, and Mike formed various other companies within the group. One was Berowell Management Ltd, which had become the legal owner of most of the aircraft from the summer of 1984, no doubt for tax reasons.

The financial problems had other repercussions, as on 31 August Metropolitan Airways ceased operations and went into liquidation. Running an airline had not turned out as simple as it originally appeared. On liquidation, Short 330 G-METO was hastily re-registered G-METP by Brencham Leasing so as not to appear the same aircraft in the eyes of creditors. It remained parked by the hangar for some months, joining the various Aero Commanders and Jetstreams that were maintained by Glos Air.

The two Vampires were sold in November to the Caernarfon Air Museum and dismantled, although they did not depart until April 1986. However, plans still went ahead for the 1986 air show season, where, in addition to G-HUNT, the two Jet Provosts would display as the *Provost Pair*. During 1985 Mike had met up with John Bradshaw, a British Airways Concorde pilot, who owned and raced piston Provost G-AWPH. This was in a red paint scheme, very similar to G-PROV. Discussions during the autumn led to the two owners agreeing to display these two aircraft instead as the *Provost Pair* during 1986, working out a synchronised display routine between themselves. Jet Provost G-JETP would be flown as a solo display aircraft by Adrian. A publicity shoot of the *Provost Pair* by Peter March was arranged for January, and later publicity described the display as combining 'The Piston Era' and 'The Jet Era'.

Peter: In the past, publicity photos for Hunter One had been taken by the late Arthur Gibson. Having visited the hangars on a number of occasions, Eric knew of my interest in the historic jets. So, I was pleased when he asked me to take on the task. I arrived on a cold January morning when Adrian and John were ready to fly their aircraft. I was to fly aboard a Jet Ranger camera ship. My problem was that, having to fly up to about 7,000 ft to find clear sky, I felt I was suffering frostbite sitting in the open door of the helicopter. Nevertheless, I obtained some excellent pictures.

Despite the downturn in operations, there was still a Christmas party for the remaining engineers at the beginning of December, followed by a drinks party at Mike's home at Northfield near Westerham closer to Christmas. However, the main party followed on 18 January, when Hunter One invited guests to their 'Summer Party' at Patterson House. Mike's idea was to dispel any thoughts of gloom!

The New Year of 1986 found there were still no further buyers for any of the fleet, so it was business as usual. During March a smoke system was devised for G-HUNT, with diesel fuel being carried in the inboard port drop tank and then piped to the jet efflux. Similar work undertaken on G-BOOM. Mike flew G-HUNT on 31 March, when the new smoke system worked as planned. On 19 April he displayed at Dunsfold for their thirtieth anniversary event. Besides air display work, the smoke was used for a Guinness TV advert shoot at the end of April.

Adrian: For filming, HUNT and BOOM had to overtake the Learjet camera-ship, with smoke on. Flying over the Learjet resulted in excessive wake turbulence for the Lear and the smoke breaking up too quickly – it was meant to be clearly seen. So, it was necessary to fly below the Learjet to obtain the necessary effect. The hours of flying resulted in only a few seconds' worth of flying sequence in the advert. Added to this our hard work wasn't seen by home audiences, as the advert was only used in Ireland and the USA!

Consideration was given to undertaking restoration work on G-VIXN, but at the beginning of February the CAA advised Brencham that the fighter was outside of the range of aircraft that they would allow on a Permit to Fly. The CAA regarded it as too complex an aircraft, so it was decided to suspend any further work. Added to this, it was discovered that the original dealer did not have ownership of the Vixen, as he had led Mike to believe, and so Brencham did not have legal ownership. It remained parked outside, along with Meteor VZ638, which was now in primer paint scheme as G-JETM. Sea Hawk G-JETH (2nd), now also painted red, was towed across the airfield at the end of February for use as a gate guard by the Bournemouth Flying Club. G-VIXN was eventually sold in April 1990 for £10,000, at a Christie's auction. It was purchased by Peter Vallance for his Gatwick collection, with profits going to the BBC's Children in Need appeal.

Towards the end of 1985 plans were made to operate G-PROV alongside privately owned Provost G-AWPH for the 1986 air display season. Training flights were undertaken the following spring but the idea of the *Provost Pair* did not work out. (Peter R March)

HUNTER ONE SUMMER PARTY

THE HUNTER ONE TEAM REQUEST
THE COMPANY OF

Mr + M

AT OUR SUMMER PARTY ON SATURDAY 18th JANUARY 1986
AT PATTERSON HOUSE BIGGIN HILL AIRPORT
FROM 7.30pm

SUMMER DRESS - *UNIQUE PRIZES FOR THE MOST "SUMMERY" DRESS*

RSVP to

Debbie Robinson
Patterson House
Biggin Hill Airport
Biggin Hill, Kent. TN16 3BN

The winter of 1985 proved a gloomy time for Hunter One, following the announcement that part of the fleet was to be sold off. Mike Carlton wanted everyone to know that operations would continue, and so arranged a 'summer party' at Patterson House in the middle of winter.

One of the aircraft affected by the cut backs was Sea Hawk G-SEAH. This had acquired a red coat of paint during 1985 and was loaned to the Bournemouth Flying Club as a gate guard from the following spring. (MHP)

During 1985 Bournemouth Airport had provided a large grass over-run area at the end of the airport's main runway. It was first put to the test by Hunter One! On the evening of 29 April 1986 Mike had taken a friend for a jolly in G-BOOM. Showing off on his return, he approached too fast and high, resulting in a late, fast landing which ended with the Hunter being parked on the grass run-off area.

> *Eric:* I was watching Mike's landing, and realised there was a problem when he disappeared from view. The tower advised that BOOM was off the end of the runway, but that both pilots had got out OK. When I arrived they were both sheepishly looking at the red-hot brakes. Mike explained that he realised he was too fast, so applied maximum brake on landing, and then he froze (Hunter brakes 'fade' rapidly in these circumstances). This meant he completely forgot to use the tail-braking parachute! Pilots again!
>
> BOOM was starting to settle into the soft grass, and the tower wanted it out of the way a.s.a.p. I realised that if it was left until the following day, a Sunday, there would no doubt be difficulty in hiring a large enough crane to lift the seven-ton aircraft. So, with engineers David and Ivan, I found a large sheet of heavyweight plywood, which we cut into three. We dug the grass and earth away as much as we could to get the boards under the wheels, which had now sunk further into the ground. Realising that we would not be able to tow the Hunter onto the boards, drastic action was called for. So, I called the tower and advised them that we were going to power up the aircraft and drive it onto the boards. I climbed in, started the engine and slowly opened the throttle, realising that by now the jet pipe was very close to the ground. As I applied further power, I had memories of the disaster

movie Airport, likening myself to Petroni, the Chief Engineer, who tried to move a 707 during a blizzard. I was almost up to maximum revs before BOOM moved, and then it all happened very quickly. Up she came, the chocks were thrown in either side of the wheels and I shut down. By now it was dark, but the Hunter was safe for the night! The following morning we laid out more boards which enabled BOOM to be pulled onto the peri-track and then returned to the hangar. The Hunter was back on the flight line by mid-day.

You may ask "What of the two pilots?" Long gone, of course (typical pilots), they waved us goodbye on the Saturday evening, headed for the nearest bar and said "See you all tomorrow", knowing full well that we engineers would of course fix it!!

May 1986 was a busy month. On 14 May the TVS film crew were busy at Bournemouth prior to the departure of G-HUNT, G-BOOM and G-JETP to Biggin Hill Air Fair, with G-PROV following the next day. All went well at the Air Fair over the weekend of 17-18 May, with the aircraft returning to Bournemouth for more TV filming. G-PROV then made a solo dash across the Channel for the

Hunter One 1986 logo

Caen Air Show on 24 May, returning the following day. For the TVS Air Show South at Bournemouth on 31 May-1 June two Hunters and two Jet Provosts were flown, with G-LOSM on static display whilst awaiting its Permit to Fly. Planned involvement of the *Provost Pair* did not work out, as the pilots could not agree on operating procedures. So, G-AWPH and G-PROV did not make their anticipated flying début at the show. In the early days of planning it was also the intention that the Sea Vixen would also participate in the show, but, as already mentioned, this idea remained grounded. All this was rather unfortunate, as Wilf Hardy's show programme cover depicted G-HUNT with the *Provost Pair* and a red Sea Vixen, all flying in formation. Meteor G-JETM and Sea Hawk WM983 were both on static display; G-SEAH was still being worked on in the hangar with the intention of returning to the air by the summer of 1987. The 1986 TVS Show marked the twenty-first anniversary of the *Red Arrows*, as well as being their 2,000th display – something that Hunter One looked up to.

G-LOSM's Permit to Fly was re-issued on 10 June, enabling the Meteor to make another long-range sortie to the continent. This time it was to the Swiss Air Force base at Sion for the air display on 14-15 June – again, all went well. Other air show work saw G-HUNT display at the BAe factories at Hamble and Dunsfold on 14 June; G-LOSM at Woodford's RAFA display, followed by Hawarden on 28 June; G-PROV at the Fighter Meet at North Weald on 28-29 June; G-HUNT at White Waltham air display on 29 June; G-HUNT and G-PROV at Lakenheath open day on 12-13 July, where G-HUNT needed engine attention prior to displaying at Middle Wallop the second day; G-HUNT and G-PROV at RAF Brawdy's mid-week open day on 24 July; G-HUNT, G-JETP and G-PROV at Goodwood on 26 July; G-HUNT at Alconbury Air Tattoo and G-LOSM to Yeovilton's Air Day, both on 27 July; G-BOOM at Exeter air display and Hullavington on 2-3 August, with an overseas trip to Gütersloh, Westphalia, and Soest, Holland, on 10 August, with G-HUNT and G-LOSM off to Eindhoven's open day on 20 August. G-PROV made a short trip to Weymouth Carnival on 20 August, and it joined G-JETP at West Malling on 25 August for the Great Warbirds Display. None of this activity gave any indication of the impending dramatic ending of operations.

In August 1986 Brencham sold 67% of its holding in Glos Air to Eagle Beechcraft, who were moving their Beech Aircraft Agency to Bournemouth. The intention was to raise cash for future business operations, including the expansion of Brencham Historic Aircraft, plus funding the newly completed Hangar 332. Unfortunately, events did not work out as planned.

8

End of Operations?

An unforeseen tragedy was to bring about the end of the Hunter One operations. Mike Carlton and his wife Kathy took a summer holiday in East Africa. On 31 August 1986 they were both killed when the Republic Seabee seaplane in which they were passengers crashed and burnt out, shortly after take-off near the Kariba Dam in Zimbabwe. The news was a tremendous shock to Brencham people in London, Bournemouth and Biggin Hill.

With the 1986 display season nearly over, G-HUNT and G-LOSM attended Eindhoven's open day on 20 September, after which most of the Hunter One fleet returned to Biggin Hill. With no-one at the management controls, operations slowed down pending a review of the future by the 'money' people in London. Tied in with the hangar sale to Eagle Beechcraft, plans were already in hand for some operations to be moved from Bournemouth to the former Norman Bailey hangar at Eastleigh. Work had started on painting and repairing the hangar, but then everyone was called to a meeting at Biggin Hill on 19 September. Here the financiers dealing with Brencham's affairs explained that there was no money available for further operations. In fact there was no money for immediate settlement of outstanding wages, creditors or hangarage fees! Without Mike's driving force Hunter One would probably cease to exist, and the fleet would be put up for sale. Many people had been aware that finances were tight, but this was a total shock to all. The Bournemouth team returned home to ponder their future, with Eric concerned as to the fate of the aircraft. Eagle Beechcraft were now taking over the Glos Air hangars and wanted the aircraft moved out – but where to? Luckily, Eric had a contact at Lovaux on the other side of the airfield at Bournemouth. He had known David Lock from his Airwork days, and David agreed to the short-term storage of the aircraft, initially only G-HUNT, G-JETM and G-JETP. The large stock of aircraft spares were gradually sold off, enabling all wages, outstanding bills and creditors to be paid before Hunter One finally shut up shop, debt free.

Mike and Kathy Carlton's funeral was held in their hometown of Westerham on 15 September. This was followed by a memorial service held at Patterson House at Biggin Hill on 31 October, with Meteor G-LOSM as a backdrop. The service was conducted by Group Captain W.E. Mantle, the chaplain from RAF

Biggin Hill, with a tribute paid by Khalid Aziz, who was an aviation enthusiast as well as TV personality. Adrian Gjertsen read the well-known poem 'High Flight', and after the service a flypast was provided by G-HUNT, G-BOOM and G-PROV. As well as many friends, flying visitors included Red Arrow Hawk XX306, Harrier GR.3 XW763, Spitfire AR213 and Corsair NX1337A. For a time the airworthy aircraft remained at Biggin Hill, although G-HUNT returned to Bournemouth on 13 December. Peter March commented on Mike's death, 'The aviation scene has been robbed of yet another of its dynamic and colourful individuals who will be sadly missed by his many colleagues and friends'.

The weeks dragged by, and there was still no definite news of what was going to happen to the aircraft. As his fellow engineers had left, Eric would visit Bournemouth to keep an eye on the aircraft. One job came along from Lovaux, as they were involved with Hunters. Back in September 1985 they had been awarded the contract to overhaul the Yeovilton-based FRADU Hunters. At the end of 1986 Eric received a request for help from Davis Lock. They had purchased Hunter T.7 XL621 from RAE Bedford, but its Avon engine had been removed as unserviceable. So, in return for the favour of a free supply of hangar space, Eric organised a working party to go to Bedford, taking along G-HUNT's spare engine, to install it in XL621 and air-ferry it back. Now registered G-BNCX, the Hunter was flown to Bournemouth by Adrian on 30 January 1987. After all that it was only used as a spares source by Lovaux. Another arrival at Bournemouth at the end of 1986 was Venom G-GONE, owned by Philip Meeson, the Managing Director of local airline company Channel Express. At the time Philip originally commenced negotiations to purchase the Venom, he anticipated it being maintained by Hunter One, but in the event it temporarily ended up hangared with Glos Air.

Although not part of the Jet Heritage fleet, Venom G-GONE resided in Hangar 600 and frequently visited the same shows as the Jet Heritage aircraft. Its naval colour scheme betrays the fact that it was a former Swiss Air Force machine. (JHL collection)

Eric: In a sense, I was now living from hand to mouth, as I no longer had a regular income. There were the odd jobs for Lovaux, plus dealing with some of the left over spares. On 30 March 1987 I officially ceased to be employed by Hunter One, but was then taken on part-time by David Lock to work in an advisory and spares support role. This enabled me to keep an eye on the stored aircraft. Adrian was not quite in the same boat as me, but he kept his hand in by test-flying the FRADU Hunters after overhaul at Bournemouth.

On 7 February 1987 G-BOOM, G-LOSM and G-PROV were all flown back to Bournemouth from Biggin Hill. This was prior to the official announcement in March that the Hunter One collection was to be sold at auction, and arrangements were made by Christie's of London for this to take place at Bournemouth on 20 July. The Historic Aircraft and Aeronautical Collectables Auction was delayed and did not take place until 1 October, when it was held in one of the former British Aerospace hangars, which provided plenty of space. Eric was hopeful that the Hunters would sell for about £250,000 each, saying, 'Mike was a marvellously generous man. It will need another wealthy enthusiast to operate these aircraft'.

Although the main purpose of the auction was to dispose of the Hunter One collection, the opportunity was taken to include other aircraft. These included a partly restored Fairey Battle, an airworthy Spitfire PR.XI, and the non-flying replica aircraft formerly displayed by Leisure Sport at Thorpe Park. The auction commenced with the usual books, paintings and other memorabilia. In fact, the local media were more interested in a fifty-year-old tin of Crosse & Blackwell Tomato Soup commemorating the 1934 England to Melbourne Air Race, which went for £800, than the aircraft. Most of the Jet Heritage aircraft failed to reach their reserves, the highest bids being:

Prior to the Christie's auction of October 1987, the former Hunter One aircraft were available for inspection by prospective purchasers. There was plenty of room, as the auction took place in the former BAC One-Eleven production hangars at Bournemouth. (Ivan Gannicott)

As well as the Hunter One fleet, the auction also consisted of additional aircraft of which a Spitfire and Battle can be seen. Besides the aircraft, there was a lot of aviation memorabilia gathered together for the sale. (Ivan Gannicott)

Hunter F.51	G-HUNT	£150,000
Hunter T.7	G-BOOM	£100,000
Jet Provost T.52	G-JETP	£55,000
Jet Provost T.52	G-PROV	£60,000
Meteor T.7	G-JETM/VZ638	£1,300
Meteor NF.11	G-LOSM/WM161	£22,000
Sea Hawk FB.3	G-SEAH	£5,500
Sea Hawk FGA.6	WM983	£1,500
Battle I	R3950	£38,000
Camel Replica	B7270	£32,000
Spitfire PR.XI	PL983	£300,000

Only G-HUNT was sold on the afternoon, along with an amount of spares. When totalled up afterwards, it was found that the actual sales proceeds of Hunter One aircraft was less than the £300,000 paid by collector Doug Arnold for the Spitfire.

> *Eric:* I was greatly disappointed with the auction, as I felt the aircraft had been sold at far too low a price. Perhaps it wasn't the best time or way to dispose of the fleet.

Various collectors attended, with the auctioneer spending some time on the memorabilia lots prior to coming to the aircraft. Unfortunately, there was not as much interest from warbird enthusiasts as had been hoped for, with only low bids being made for the aircraft. (Ivan Gannicott)

Following its sale, Sea Hawk G-JETH departs for its new home at the Vallance Byways Collection near Gatwick Airport. In fact it had not reached its reserve price at the auction, being sold a few weeks later. (JHL collection)

Flagship Hunter G-HUNT was another of the fleet sold after the auction. Initially thought to have been sold to an Irish buyer, the transaction fell through and G-HUNT was resold to a US enthusiast. It is seen in its packing case at Bournemouth prior to departure. (JHL collection)

Having 'lost' G-HUNT, Eric Hayward was pleased that it ended up in suitable hands. Restored by Jim Robinson of Texas, the Hunter was repainted to resemble Neville Duke's WB188. Brian Henwood undertook the initial test-flying from Houston Hobby Airport. (JHL collection)

On the evening of the auction an amusing event occurred when I was contacted by the purchaser of HUNT. He was an agent acting on behalf of an Irish businessman who had instructed him to purchase BOOM – the two-seater. The agent was now feeling very uncomfortable on realising that he had spent £150,000 of the businessman's money on the fighter version, so he sought my advice. It also turned out that the police were interested in the transaction, wondering why an Irish person had purchased a fighter – remember these were the times of the Troubles. The problem was solved when I found a replacement buyer for HUNT. The Irish businessman never got his high-speed executive jet.

After its abortive sale to Ireland, G-HUNT was sold to Jim Robinson of Houston Hobby Airport, Texas, eventually being shipped out there early the following year. Jim had originally shown interest back in June, when he visited Bournemouth. Then Eric flew to Texas in July to discuss Jim's operations there and the support he would need for a Hunter. Later, Jet Provosts G-JETP and G-PROV were sold to two businessmen brothers, Ian and Dougal Craig-Wood, who had only intended to buy one, but then decided to have one each. They negotiated a price below the auction reserve, also purchasing G-BOOM and G-LOSM at a later date as both had failed to reach their reserve price. Red Sea Hawk G-JETH (2nd) was sold after the auction, going to the Vallance Byways Collection near Gatwick. Finally, Meteor G-JETM was sold after the auction to Aces High at North Weald.

As mentioned, the one aircraft to be 'lost' from the UK was G-HUNT. After the auction it was given a final engine run by Eric, dismantled for packing, and departed by sea for the USA in November 1987. Eric had already visited Houston to see Jim's 'Combat Jets Museum', which included two F-86 Sabres, an F-104 Starfighter and a T-33. G-HUNT was re-assembled in the following spring, and Eric was asked back to supervise its initial operations. Unfortunately, he had an operation of another kind to attend to – a hernia repair carried out at Salisbury Hospital. This meant he did not get to Texas until the middle of June, just in time for initial engine runs. On 23 July G-HUNT, now registered N611JR, made its first flight in the States with Jim Robinson at the controls. A repaint was undertaken which included serial WB188, so that it now represented Neville Duke's record-breaking aircraft. Brian Henwood flew over from England at the end of July to undertake some further test-flying before the Hunter was displayed at Oshkosh. At the end of its flying days the aircraft was presented to the Oshkosh Museum by Jim Robinson.

This marked the end of Hunter One; at that time no one was aware of the phoenix waiting to rise.

9

Phoenix Rises

During the months prior to the Christie's auction, Adrian had been considering future opportunities. While flying for Mike he had been bitten by the historic jet bug, and he wanted some sort of operation to continue. He was able to proceed further with his plans at the end of 1987, once the outcome of the auction was known.

> *Adrian:* Following the auction Eric introduced me to the Craig-Wood brothers. I proposed that it would be best to keep the collection together, and to re-establish a flying museum of early British jet aircraft. At the time I was unaware of their great interest in aviation, and they were grateful that I offered my services to them. Agreement was later reached to form a new company, with the aim of preserving the historic jets. Better to keep them flying, so as to keep at bay deterioration that would plague the aircraft if they were merely static exhibits. The project's eventual success was a tribute to the far-sightedness of the new owners and investors.

Following the Christie's auction, Jet Provosts G-JETP and G-PROV were moved to the Interair hangar at Bournemouth for a further period of storage during the autumn of 1987. They had been purchased by the Craig-Woods, and it was the new owners' intention to move them to Biggin Hill for overhaul. The brothers were in the computer business, but were also former UAS student pilots and long-standing aviation enthusiasts. They already had a Vampire T.11 under restoration at Cranfield. Finding Adrian's plans most interesting, they asked if the new venture could include their Jet Provosts. No problem! Within a short space of time the brothers decided to add the unsold Hunter G-BOOM and Meteor G-LOSM, which were initially registered to LGH Aviation. The purchase was completed in January 1988, with the aid of financial support from an overseas trust. An engineer would be needed to look after the fleet. Adrian had to look no further than Eric Hayward who was more than happy to be involved and be able to take on some of the former Hunter One engineers. Now for somewhere to keep the aircraft!

The original Glos Air hangar had been used by Eagle Beechcraft, but their business had not flourished as expected. The downturn in the economy at the

There was activity around the former Glos Air hangar in the autumn of 1988 with the arrival of two Hunter T.7s from RAF Cosford. These were transported and assembled by Lovaux assisted by the Hunter One engineers. (JHL collection)

The two Hunters were reassembled by early 1989. At the time the future plans for the aircraft were kept under wraps. XL572 also carries its Cosford maintenance serial 8834M, and Eric later added its civil registration G-HNTR. (MHP)

time adversely affected the business-aircraft market. Discussions were held at the beginning of 1988 between Ian and Douglas and Eagle Beechcraft, their lease having only a few months to run. With interest shown in taking over the hangar, Eagle Beechcraft set a date of August when they would move out. On 18 August a new lease was signed, which enabled Hangar 600 to house the historic jets once again, the initial four being G-BOOM, G-LOSM, G-JETP and G-PROV. Added to these were Sea Hawk G-SEAH, which had been bought by Adrian and registered to a company he had set up, Sark International Airways. This was regarded as a longer-term restoration project. More immediate was the acquisition of further Hunters from the RAF. A number of Hunter T.7s were surplus to the requirements of No.2 SoTT at Cosford, and were offered for sale. The aircraft were surveyed in July, and the opportunity was taken to buy three: XL572, 576 and 617. XL572 and 617 were moved by road to Bournemouth in October, but XL576 was sold on for use as an instructional airframe. Despite having the aircraft at Bournemouth, the new team was unable to organise anything for the expanded TVS Air Show South on 4-5 June, although Philip Meeson flew his Venom.

Now having aircraft to work on, Eric was officially taken on as Engineering Director from 1 November (although at the time he was enjoying a pre-planned holiday in Kenya). Eric always preferred to be known as Chief Engineer, in the belief that anyone could aspire to the title of Engineering Director and a nameplate on one's office door. Chief Engineer indicated that one had worked their way up through the ranks to achieve a position of respect with the relevant knowledge that went with it. As mentioned previously, Mike Carlton had been in negotiations with Flint College as early as 1984 for the acquisition of their unique Swift F.7 XF114. Further negotiations by the new team resulted in the Swift being purchased in November 1988, dismantled in December and moved to Bournemouth in January 1989.

10

Jet Heritage

To run future operations, a new company, Jet Heritage Ltd, was formed in the summer of 1988, thus securing the future of the former Hunter One collection. Ian and Douglas arranged the majority of the finance through trust funds, with Adrian appointed as Operations Director & Chief Pilot and Eric as Engineering Director. As well as maintaining their jet fleet, the company would also undertake maintenance and overhaul work on behalf of third parties. Adrian realised that by keeping the aircraft flying there would be a pool of pilots remaining current on the types, also preventing deterioration that would occur if they remained grounded.

A press launch for Jet Heritage was held at Bournemouth on 8 February 1989, covered by the national as well as the local papers, followed by VIP launch on 21 February. Visitors were welcomed to 'The Finest Collection of Flying Historic British Jet Aircraft', with the static Swift taking pride of place. This was backed up by the Jet Heritage brochure promoting the aim of 'Maintaining the Legend' and 'Preserving the finest of jet aviation heritage in the finest tradition'.

> *Adrian:* At the launch I stated that Jet Heritage was proud to be carrying on the work and aspirations of the Hunter One team – to preserve Britain's fine military jet tradition. Our aims were simple: to restore and maintain, in full flying condition, as many examples of our classic fighting jets as possible, and to display them to the public. We would like to express our gratitude for the idealism of those without whose generous support a part of our aviation heritage would undoubtedly have been lost forever. The hard work of the past few months had come to fruition. Without the idealism and generous support of the Craig-Wood brothers, part of our aviation heritage would have been lost. At the time only G-PROV was airworthy, and I gave a demonstration for the invited guests. I explained that G-JETP was being worked on in the hangar and would soon join G-PROV, both being used to check out pilots. Next in line would be G-BOOM.

The fleet would be available for film work in addition to air-show appearances. In addition, Jet Heritage intended to attract further historic jets to Bournemouth for restoration, hangarage and maintenance. A major bonus with Swift XF114 was that it came with a full set of manufacturer's blueprints and drawings, so

Jet Heritage was launched with much publicity at Bournemouth on 8 February 1989. A number of the Hunter One aircraft were included in the fleet, along with a Swift, which was regarded as the star item. A new dawn for classic British fighter preservation had arrived. (Peter R March)

The Swift was a type that had been sought for a number of years, with Mike Carlton starting negotiations. In the end it was Jet Heritage that completed the transaction. XF114 is seen at Bournemouth on the occasion of the Jet Heritage VIP Launch Day in February 1989. (MHP)

A replacement Hunter fighter was sought in 1989 to operate alongside G-BOOM. This resulted in XE677 being acquired from Loughborough University, where it had been well cared for in its role as an instructional airframe. (JHL collection)

increasing the possibility of restoring the aircraft to flying condition. However, it was estimated that it would need at least four years of work by Jet Heritage before being able to fly. As a replacement machine for the college, Jet Provost XR658 had been purchased from the RAF in June 1988 (replacing XR654) and moved from storage at Wroughton to Bournemouth in September. The airframe had been overstressed during a heavy landing, but would be suitable for ground instruction. An old, but usable, Viper engine was fitted to enable the Jet Provost to undertake engine runs. The rest of the fleet was made up of G-BOOM, G-LOSM, G-JETP, G-PROV, G-SEAH, Hunters XL572/617, with Vampire XH328 to come. The majority were owned by Hunter Wing Ltd, a Jersey-registered company, although the UK operation was set up as Jet Heritage Ltd.

Another fighter now maintained by Jet Heritage was Philip Meeson's Venom G-GONE, which required an engine change during March. Philip also purchased a Dragon Rapide G-AGSH, which arrived by air early in 1989 in Alderney Airlines colours. Although not a jet, Jet Heritage was happy to house the biplane airliner. Philip favoured this more sedate form of flying, although G-GONE was flown at the North Weald show in May. The Venom was frequently flown by John Davies, a BAe test pilot, so later in the summer Philip sold him a 50% share in the fighter.

Eric: The Rapide's registration rang a bell, although I was put off by its being painted with Alderney Airlines titles. It had a family association, as G-AGSH was the BEA aircraft my father had flown in from Croydon to Jersey for a holiday in 1947, and the registration GSH also being his initials – George Stewart Hayward. Some sixteen years later I got married and we travelled to the Scilly Islands for our honeymoon, where the final leg was by Dragon Rapide from St Just. Our return flight to the mainland was by G-AGSH. I related my family involvement to Philip, whereupon he asked if my father was still alive. When I confirmed that he was, Philip said "Bring him along, and we shall give him a repeat flight". So, on 15 April 1989, we took off with my father sitting in the same seat in the same aircraft that he had occupied over forty years before. Philip gave us a thirty-minute flight along the coast, which my father enjoyed immensely.

The new hangar 332, which had been built adjacent to 600, was now used by Super-X for the production of display simulators, which were becoming popular attractions at air shows. Their *Red Arrows* one was most popular, but they also devised one for a Tornado. This brought the visit of two Tornado GR.1's on 29 March for the simulator's launch – although their crews were more interested in the contents of the Jet Heritage hangars.

The Swift, now registered G-SWIF, was moved inside the hangar during May so that a full examination of the airframe could be undertaken. Eric considered that it was built more like a warship than a fighter, and that perhaps it had been designed by Vickers Shipbuilders instead of Vickers-Supermarine. Initially, restoration work was concentrated on its wings. The summer months proved a busy period, with the acquisition of further aircraft. Jet Provost T.5 prototype XS231 fuselage was purchased from the RAF at Scampton for spares use, arriving on 4 July. G-BOOM flew under Jet Heritage's ownership on 16 August – the first time for 2½ years, as its Permit to Fly had expired. It was then time to find another Hunter fighter. Loughborough's University of Technology had been using Hunter F.4 XE677 as an instructional airframe for a number of years. Now surplus to their requirements, it had been moved to nearby East Kirkby for storage. So, on 4 August a visit was made to East Kirkby in a Robinson R-22 to check on the state of the aircraft.

Eric: I learnt that the University was going to sell their Hunter by tender, with bids required by 11 August. Having satisfied myself that the airframe was in good condition, I contacted the University's Finance Department and established that they considered £5,000 would be a good sum to realise. So, Jet Heritage put in a successful bid of £5,200, which they seemed most satisfied with. I was also pleased, knowing that Jet Heritage had acquired the Hunter for such a reasonable price. So, Operation 'Get Old 677 Back into the Air' was launched, no one realising at the time that it would become the flagship of our fleet. I also found that it was one of the few F.4s fitted with the wing leading-edge extensions, common to the F.6.

The Hunter moved to Bournemouth by road on 8 September, and was registered G-HHUN to Hunter Wing in October. The aircraft needed a lengthy rebuild, as, although it was low on airframe hours, it was a long way behind on current Hunter airframe modifications. In fact, it was to be four years before power returned to the aircraft. Work undertaken included the fitting of a braking parachute, upgrade of the radio equipment and the addition of an electric starting system as fitted to G-BOOM.

A visit by Jet Ranger was made to Cardiff on 21 September to view the aircraft being sold in their museum's auction, but there was nothing suitable for Jet Heritage. At the end of the month a visit was made to Woking to view Gnat T.1 XM697, which was on display with the local ATC cadets. This proved to be an early production aircraft which Eric did not consider worth bothering about. In the event Ian and Douglas purchased it on a whim, as at the time Gnats were the 'in' warbird, and it was felt that anything could be restored. It moved to Bournemouth

XE677 was dismantled and transported to Bournemouth in September 1989. The airframe was in a sound condition, but was not up to the latest modification standard of RAF Hunters. Luckily, an ample supply of RAF Air Publication manuals was held for use by the engineers. (JHL collection)

Venom G-GONE demonstrates its engine's cartridge starting system. Jet Heritage had already fitted its Hunters with an electric starting system which, although less spectacular, was much easier for the ground crews. (JHL collection)

Though it was now part of the Jet Heritage fleet, restoration work on Sea Hawk G-SEAH only proceeded as time allowed. Other projects were regarded as higher priority, and so it was rare to see the engineers working on the Sea Hawk. (Peter R March)

on 21 November and was registered G-NAAT, although Eric considered it only suitable for spare parts. A further Sea Hawk (8151M) was purchased, arriving by road from a collector at Radstock on 24 October. Hardly any restoration work had been undertaken, so it was stored to await its turn for rebuilding.

Due to the workload in restoring the various aircraft, there was little time during the summer to send aircraft to air shows. G-JETP visited the regular Dunsfold's Families Day on 1 July, but other appearances were left to G-GONE. The Venom displayed at Hatfield's families day on 1 July and then Yeovilton's air day on 29 July.

When the Brencham Group ceased trading, part of the set-up included the hangars at Biggin Hill. When these were sold in 1989, there was also the gate guard Hunter to consider. It was not needed by anyone at Biggin, and so was 'reclaimed' by Jet Heritage and returned to Bournemouth at the beginning of November. The final aircraft activity in the year related to Hunter XL617 (registered G-HHNT July), which was sold unrestored to an aircraft dealer in the USA. This involved it being dismantled during November and fitted into a crate for shipping across the Atlantic. It was registered N617NL to Northern Lights Inc in Canada and painted in smart blue colour, and later sold to Steve Appleton USA in 1999 and repainted red, similar to G-BOOM.

During his lifetime Mike Carlton had supported a number of aviation causes, so it was no surprise when he agreed to assist the New Milton ATC cadets in fundraising for a new HQ. In the summer of 1985 the cadets had held a sponsored tug of G-HUNT around Bournemouth perimeter track. By 1989 they had raised sufficient money to fit out their new HQ building, which had been provided by Esso Petroleum. In recognition of Mike's assistance, the ATC Committee agreed to name the HQ 'The Michael Carlton Building', and it was opened by Air Marshal Sir John Curtiss on 21 October 1989.

1990 began with severe storms, which damaged Hangar 600's roof on both 25 January and 26 February. Pending full repairs temporary sheeting was fitted, something that was not within the normal scope of airframe engineers!

At the end of January 1990 Eric visited Fort Wayne, USA, to undertake a survey on F-86A Sabre NX178. British collector Robert Horne showed some interest, and he sought Jet Heritage's expertise on the fighter's condition. Another collector, Frank Hackett-Jones, sought Jet Heritage's help, which resulted in the purchase of Gnat T.1 XR537 on his behalf in March. Frank was the director of a business-aircraft leasing company in Guernsey. The Gnat was one of the lots at Sotheby's auction held at RAF Cosford on 9 March. The auction catalogue indicated a price guide of £45-60,000 but Frank was so desperate to acquire a flyable Gnat that he instructed Adrian to buy one at any price. He did – paying £105,000! (A US dealer paid £122,000 for his.) The Gnat, still in *Red Arrows* colours, arrived by road on 5 April, in one piece, carried in a special cradle for support. It was registered G-NATY in June. A former ground-running instructional airframe, it was found to be in excellent condition, and Eric

Jet Heritage undertook overhaul work for other owners, hence the arrival of former Red Arrows Gnat XR537 for Frank Hackett-Jones in April 1990. It was anticipated that would be in and out of the hangars within a few months – unfortunately it was years! (JHL collection)

anticipated that it could be returned to the air within three months – six at the outside. Unfortunately, this turned out to be optimistic thinking.

Meteor G-LOSM had been built as a night fighter and, as such, only had one set of controls. The cockpit layout was almost identical to a Meteor T.7, and so plans were made to fit a second set of controls to the rear seat of G-LOSM. These were taken from Meteor T.7 WH166, whose front fuselage arrived from Scampton in April. The removal took about ten weeks, after which WH166 was returned to Scampton. Having a dual control Meteor would have been a great advantage, but unfortunately, after all the effort, the controls were never fitted. The T.7's cockpit was unpressurised, whereas the NF.14 was pressurised. This made the fitting of the rear seat controls much more complex than anticipated, and so they were put aside for a later day. G-LOSM was repainted after a major inspection early the following year, resulting in its 141 Sqd. marking being removed. Jet Provost XR658 was completed by late spring, moving by road to Flint College on 1 June 1990, and was officially handed over on 7 June for ground instructional use. The spares ship fuselage of XS231 was sold to Bruntingthorpe and departed in the autumn. A visit was made to Cranfield at the end of August to view the Craig-Woods brothers' partly restored Vampire T.11 XH328. Then a visit was made to Duxford in September to view CT-33 Silver Star N33VC/'54-21261' as a possible new project, but it was considered unsuitable.

An unexpected official approach was received from Rolls-Royce at Hucknell in March 1990. They were using two Derwent engines on their test beds and often had difficulty in starting them. Eric flew up, and was given an interesting tour of their factory before undertaking a survey of the engines concerned. Finding they were Mk.8 engines, which were known for poor starting, Eric was then contracted to supply a working party and parts to modify them to Mk.9, which was soon done. As a result, Jet Heritage received an appreciative letter from Rolls-Royce on completion. This pleased Eric, as it showed the high regard that both the CAA and Rolls-Royce had for Jet Heritage's work.

Further engines were acquired from various sources. The visit to Rolls-Royce at Hucknall produced a Nene and Derwent; then an Orpheus came from Biggin Hill in April, with another coming from Bridport ATC in May; eight Derwents also arrived from France in October. The Orpheus was in good condition, but had no logbook. It was stripped and rebuilt under Rolls-Royce supervision, and later fitted to Gnat XR537 as a serviceable unit. Further Derwents were obtained from a Portsmouth scrap dealer, proving to be examples of the 'snow blower' version. These had been modified by the RAF to blow snow off runways. Eric knew they could be converted back for use in G-LOSM, but there was no paperwork supplied with them. Armed with the engine numbers, Eric visited Rolls-Royce at East Kilbride to obtain duplicate paperwork. Unfortunately, all such documentation had been thrown out when Rolls-Royce went 'bust' in the 1970s. Luckily, some of the employees had rescued the filing cabinets containing such paperwork, and Eric found Log Cards for two of his engines. This was good enough for CAA requirements.

Relations with the CAA had been built up over the years, starting when G-HUNT's Permit to Fly was signed off. On the engineering side their inspectors got used to the idea of dealing with Permit to Fly paperwork for a fighter, not a C of A for a Boeing. Some of the inspectors became well-known faces, but on occasions the CAA would send down a 'new boy'. They often had not realised that Mach One fighters were part of the job description, so Eric had to enlighten them.

Adrian: During my time at Jet Heritage I became a CAA display evaluator, which involved signing off display pilots. In Britain we had more authorised display pilots than the rest of Europe, as there were more displays here than in the rest of Europe combined. From Jet Heritage's point of view these displays were used to show off the aircraft, not the pilot. The routine was on the cautious side, and although the Hunter could use up to 6g, the Meteor NF, Sabre and Vampire Trainer were limited to 4g.

Engineers attending G-BOOM at Manchester following its emergency landing in May 1990. This had resulted in its port-undercarriage leg breaking; a replacement was flown up from Bournemouth for fitting by the Jet Heritage engineers. (JHL collection)

Under Eric as Chief Engineer, there were now more mechanics working on the fleet. 1990 was the first season when Jet Heritage was able to fully display its aircraft. G-BOOM, in company with G-GONE, flew north for the Barton Air Show on 13 May. During the display G-BOOM encountered elevator problems and diverted to Manchester Ringway, where it made a heavy landing. During the landing an undercarriage leg casting cracked, requiring a replacement to be taken up from Bournemouth to enable the Hunter to be flown back two days later. At the local Bournemouth Air '90 Show on 25-27 August the Jet Heritage fleet was on static display. Originally it was intended that G-GONE take part in the flying, but in the event it was G-LOSM that flew.

After months of rumour, another warbird operator arrived at Bournemouth in the summer. Doug Arnold moved the majority of his Second World War fleet into his newly built hangar on the north side of the airport. The fleet were not very active, but it was different to see the likes of Spitfire, Mustang and Corsair in the circuit.

The empty hulk of former Danish Air Force Hunter F.51 E-402 arrived from Lovaux in December 1990 for storage on behalf of a private owner, not to become a Jet Heritage aircraft. However, it proved to be another of the Hunters surveyed by Eric back in 1975. Parts of Vampire XH328 arrived by road from Cranfield in December and, being regarded as a basket case, remained in storage. As with XM697, this had been a spur of the moment purchase, as the airframe proved to be in a far from ideal condition. Provost T.1 XF877 (G-AWVF) visited in November and, although not a jet fighter, was purchased by Ian and Douglas in December. As with Hunter One's plans back in 1985, the idea was to pair it with one of the Jet Provosts at displays.

A view of the Jet Heritage apron in the spring of 1990. On the left is the gate guard Hunter which had been retrieved from Biggin Hill, whilst on the right is Jet Provost XR658, destined for Flint College. The Devon was a visitor. (MHP)

11

Raising Public Awareness

In 1991 Eric Hayward renewed his Swiss connections, having worked there for BAe on Swiss Air Force Hunters for 3½ years. In the past the Swiss had always scrapped their aircraft on withdrawal from service. They were withdrawing large numbers of Vampires, but were now also aware of the growing warbird market. As a result, the majority of their Vampires were put up for auction at Sion. Having shown interest, Jet Heritage received a sale catalogue in January 1991, also being advised that they had been given FB.6 J-1149. Eric and Ian Craig-Wood flew to Sion for a few days in March to view the aircraft; the auction was held on 22-23 March. Having accepted the gift of the aircraft, they also agreed to buy T.55 U-1215 for the collection. After minor maintenance J-1149 flew to Bournemouth on 10 May, later being registered G-SWIS. U-1215 followed at mid-day on 28 August, flying its delivery flight in company with FB.6s J-1129/52, which then flew on to Duxford. This was the start of a number of former Swiss Air Force aircraft arriving at Bournemouth.

In January Doug Arnold visited Hangar 600 to see the set-up, and in April he purchased four former Swiss Pilatus P-3 trainers. Doug was still trying to get his fleet of aircraft flying and wanted to see how Jet Heritage performed. Another unexpected visitor was an L-39 Albatros, which turned up outside the hangar the first weekend of June. N159JC had been purchased by Rob Lamplough from Chad, and was stored at Bournemouth for five weeks on its way to North Weald.

The Jet Heritage directors decided to raise the company's profile by organising a number of aviation anniversary events, with the first being held in 1991. Jointly sponsored with Shell Aviation and GAPAN, Jet Heritage hosted a heritage event on 13 May to commemorate 'Fifty Years of British Jet Flight – a tribute to Sir Frank Whittle'. Many famous test pilots attended on the day, and Sir Frank was billed as guest of honour but, regrettably, poor health prevented him travelling over from America. Hangar 600 contained the likes of Geoffrey Bone, 'Winkle' Brown, John Cunningham, Neville Duke, John Farley, Charles McClure and many more. All were happy to raise their champagne glasses in response to Adrian's toast to Sir Frank, who sent his apologies on video. There was a short display in the morning by the Meteor, Vampire and G-BOOM, with the two Jet Provosts flying later in the afternoon. In their coverage the local press reported

Vampire FB.6 J-1149 in storage at Sion in March 1991. This was the aircraft donated to Jet Heritage by the Swiss Air Force, flying to Bournemouth in May. In the background is U-1215, which was purchased by Jet Heritage at the Sion Auction. (EGH)

Hunter gate guard mounted on its new plinth at Bournemouth. At first it still carried the Brencham 'BHAC' lettering, but by 1992 it was repainted with RAF markings and a Jet Heritage logo below the cockpit. (MHP)

that the pioneering Gloster E.28/39 was also due to fly!! There was a strong Jet Heritage presence at North Weald's Fighter Meet a few days later, with G-LOSM, G-BOOM, G-JETP, G-PROV and G-GONE all on display. The *Sunday Express*, one of the shows sponsors, ran the headline 'Boom! The return of the Hunter', referring to 111 Squadron's days at North Weald. Brian Henwood flew G-LOSM to Dunsfold's Families Day in June, but poor weather curtailed his display and delayed his flight home. The local 'Bournemouth Air '91' Show took place on 24-25 August, proving to be the last held at Bournemouth. However, this was a low-key event for Jet Heritage, with only Meteor G-LOSM being displayed. G-BOOM and G-LOSM both flew to Oostmalle in Belgium for their show on 7 September. Gate guard Hunter BHAC was erected on a plinth outside Hangar 332 at Bournemouth in June, with a plaque unveiled dedicating it as a memorial to 'Michael Richard Carlton – Aviator'.

The overhaul of Gnat XR537 reached the stage of engine runs by the spring of 1991. Unfortunately, problems arose over obtaining vital replacement parts, so, regrettably, work was suspended and the aircraft put into storage. The other Gnat XM697 was repainted from its pseudo-*Red Arrows* colours into RAF air superiority grey during the summer, but remained in storage. Visits were made to various aircraft during the year to see if they were of any use to Jet Heritage. Jet Provost G-PROV flew to Newcastle in early August to enable Eric to view Swift parts at Sunderland, but they turned out to be unsuitable for G-SWIF. At the end of the month Eric again flew to Fort Wayne to have a further look at Sabre NX178. Maintenance work had been undertaken and additional hours in the air put on the clock, enabling Eric to recommend its purchase to Robert Horne. It was dismantled for shipping to the UK via Montreal in December. Vampire U-1215 was registered G-HELV (referring to Helvetia) in September, and at the end of October Eric flew to Sion to collect some Goblin engine spares, which it urgently required. Hunter activity saw Adrian and Eric fly to Cranwell in G-LOSM in October to check over an instructional Hunter F.6 which was for sale. Back at Bournemouth, the stored Hunter XL572/G-HNTR was sold to British Aerospace for ground instruction, leaving by road in October for their Overseas Training Department at Brough. In the opposite direction, Jet Heritage received the fuselage of Hunter XG290 and a pair of Jet Provost wings from Bruntingthorpe for spares and cannibalisation. A fuller examination of Provost XF877 revealed that its airframe condition was such that it would not stand the stress of aerobatics, so it never appeared at displays but remained in a corner of the hangar. Later it was flown away to a private airstrip near Thatcham.

Jet Heritage advertised the Jet Heritage Film Service, which was managed by Adrian. This offered 'the world's largest collection of British ex-military jets' for use in film work. Shares in G-BOOM were also available, with £10,000 buying a 5% stake – not forgetting hangarage, maintenance, flying hours, fuel and landing fees! To help boost income, Jet Heritage produced sponsorship proposals, whereby, for £30,000, a complete flying programme at a show could be sponsored. Lower down the scale, an individual appearance could be provided for £2,000. There were even proposals to display company logos on one of the aircraft, or even to go as far as having an exclusive livery on one or all of the fleet. In the event, no such 're-branding' occurred. Further income was received from the Aerospace Division. Having built up a large stock of spares for the fleet, Jet Heritage was able to offer its services to other operators. Airframe parts for

After its overhaul in the spring of 1993, Vampire T.55 G-HELV initially retained its Swiss AF colours as U-1215. Upon completion of test-flying, it was repainted in a pseudo-RAF camouflage colour scheme. (MHP)

Being a non-standard airframe, Gnat XM697 was held as a possible source of spares for XR537. However, it received the civil registration G-NAAT, and in 1991 was painted in the then-current RAF fighter colour of air superiority grey. (MHP)

Jet Heritage advertised its fleet as being available for film work. Unfortunately, the hoped-for demand was not there, resulting in just a few roles in film and advertising work, so an anticipated source of income did not materialise. (JHL collection)

With a wide range of historic jets on their books, Jet Heritage needed the back-up to maintain the fleet. A vast library of RAF Air Publication manuals was acquired from various sources and proved invaluable to the engineers over the years. (JHL collection)

North-American F-84 Sabre NX178, on a test-flight over Fort Wayne, USA, in the summer of 1991. It was checked out by Eric Hayward on behalf of Richard Horne, who was negotiating to bring the historic fighter to England for the Golden Apple Trust. (EGH)

Canberra, Hunter, Jet Provost, Vampire were available, as well as a variety of engine parts. These included zero hour, packed Goblin engines.
The star acquistion at the beginning of 1992 was F-86A Sabre 48-178/NX178 which was shipped across the Atlantic. It arrived at Bournemouth on 6 January, where its spare engine was found to have been damaged during shipment. Robert Horne, who had established the Golden Apple Trust, visited on 21 January to see his new purchase, which had already been registered to the Trust as G-SABR the previous November. Its first engine run was on 24 February, with its first UK flight undertaken by Adrian on 21 May (this was also Adrian's first flight in a Sabre), followed by a display at the Biggin Hill Air Fair. On 17 June there was then an official press launch day at Bournmoth for the Sabre, which remained hangared with Jet Heritage for maintenance and display duties. There was plenty of display work for Jet Heritage, with Brian Henwood taking G-LOSM to the Middle Wallop Air Show on 9-10 May, and the majority of the fleet appearing at Eglinton later in the month. G-LOSM then appeared at Barton on 25 May, joining G-SABR at Biggin Hill on 20 June, with G-BOOM and G-LOSM both crossing the Channel to Lelystad on 29 August.

There was further activity on the Swiss Air Force front. Vampire T.55 U-1216 had been presented to the RAF in 1990, but they didn't know what to do with it. Originally they had intended to fly it at air displays, but the plan was dropped and the aircraft stored at Boscombe Down. A survey of the aircraft was undertaken by Jet Heritage on 31 January to assess its future worth. At the beginning of March Eric again visited Sion to oversee the purchase of Vampire FB.6 J-1106/HB-RVO from a private owner. He also viewed the entire package of Swiss Air Force Vampire spares which were offered to Jet Heritage. Despite the tempting offer, it had to be turned down, as the stock proved too large to pack, transport and store back at home. Later in the month Eric visited preserved Swifts at Hereford and Newark to see if there were any useful parts for G-SWIF,

Jet Heritage was given the task of re-assembling and flight-testing the Sabre. Appropriately registered G-SABR, it first flew in Adrian's hands in May 1992 (his first flight in a Sabre) and remained based with Jet Heritage until 1995. (JHL collection)

Sea Hawk WV795 was repainted into its Royal Navy colours in the spring of 1992. Never a serious restoration project, the fighter was sold to a Cypriot collector the following year, but in the end he did not bother to collect it from Jet Heritage. (MHP)

but again there were none. Sea Hawk 8151M was repainted by May, now restored as WV795, but was for static display only. The Jet Heritage engineers decided to take a day off on 27 September, treating themselves to a flight in a Beaver floatplane at Calshot.

> *Eric:* I had visited Switzerland many times in my BAe days; now I was doing it again for Jet Heritage. It was not all work, as on one visit to Altenrhein I was given an aerobatic flight over Lake Constance by Hans Kostlie in his Boeing Stearman. He was a local businessman and aviation enthusiast whom I got to know well during my time working on the Swiss Hunters.

Despite being on the look-out for further aircraft, Jet Heritage decided to enter a number of its fleet in a Sotheby's auction at Billingshurst on 19 September to 'test the water' on current prices, with Provost XF877 overflying the event to drum up interest. In most cases bidding failed to reach the reserve price, the bids being Hunter G-BOOM £70,000, Provost XF877 £15,000, Gnat G-NAAT £10,000 and Sea Hawk WV795 £2,000. This was an indication of the depressed state of the warbird market. Only Jet Provost G-JETP was sold at the time for £69,000 to Savvas Constantinides, a museum owner in Cyprus. He surprised everyone by paying cash from a purse he was carrying! However, it remained at Bournemouth for a few months longer.

Doug Arnold died in November and his warbirds moved away from Bournemouth – a loss of one of the great warbird characters and also of interesting aircraft movements. In the same month, members of the Bournemouth Heritage

To mark the fiftieth anniversary of the Meteor's first flight, G-LOSM took part in displays at Staverton and Bournemouth on 5 March 1993, in company with Martin-Baker's WL419. The Meteors are seen after landing at Bournemouth, watched by Air Traffic Staff. (JHL collection)

Both Meteors on arrival at the Jet Heritage ramp. WL419 was more representative of the early Meteor fighters – G-LOSM being the later night-fighter version. At the time WL419 was in full-time use by Martin-Baker for ejector seat trials. (JHL collection)

Transport Trust held talks with Jet Heritage to see if they could use part of the hangar to store their bus collection. However, neither side could agree terms, and the buses ended up elsewhere on the airfield.

On 5 March 1993 G-LOSM took part in a special flight to mark the fiftieth anniversary of the Gloster Meteor's first flight. It flew to Staverton in the morning, for a short display close to the former Gloster factory, in company with Martin-Baker's trials T.7 WL419, piloted by Stan Hodgkins. Both Meteors returned to

Bournemouth in the afternoon for a further display. There was a visit made to Hangar 600 on 19 March by Roly Beamont and Jimmy Dell – both noted for their Lightning test-flying. Vampire G-HELV put on a display for them.

Jet Provost G-PROV was sold on 6 March and departed by air on 12 March for Leavesden, with G-JETP departing for Cyprus on 26 May flown by Bob Thompson. Savvas Constantinides also purchased resident Sea Hawk WV795, plus a former Royal Navy Whirlwind XN385 for his collection. However, both remained stored at Bournemouth, never leaving for Cyprus. G-JETP was eventually dismantled and never flew again. Vampire G-HELV's overhaul was completed, and it was test-flown by Stan Hodgkins on 10 March, after which it visited the paint shop. It re-appeared in a pseudo-RAF camouflage scheme with serial 215. As such it flew the display circuit alongside similarly painted Meteor G-LOSM as one half of the *Heritage Pair.* This was a reference to the Meteor T.7 and Vampire T.11 flown by the RAF at displays as the *Vintage Pair.* The overhaul of XE677/G-HHUN progressed, with undercarriage and flap retractions undertaken during March. Its Avon engine was first run at the end of July, but there was still much work to undertake before the aircraft flew again. Thought was given as to its eventual colour scheme. The plane was originally in RAF camouflage, and consideration was given to retaining this, with MoD permission obtained to back in September 1990. However, some of the team wanted to see it in blue, others in black. Eric said red, and somehow this was agreed upon. No further work was undertaken on J-1149/G-SWIS, as the CAA found that there were unacceptable modifications to the airframe which would have been too expensive to resolve. So, it remained stored until eventually sold to Sir Ken Hayr and moved to New Zealand in 1999.

Bournemouth Helicopters set up business in March 1993 in a Portacabin adjacent to Hangar 600. This meant their Robinson R-22s were using Jet Heritage's apron for their training flights. Also taking up space were HS.748 airliners which were parked up between being used on nightly mail flights. More welcome were BoBMF Spitfire PR.19 PS853 and Devon VP981, which night-stopped on 15/16 May, with Sqd. Ldr Day giving a display over Bournemouth seafront in the Spitfire.

New pilots joining Adrian and Brian Henwood at Jet Heritage were John Davies and Dick Hadlow. John was a BAe production test pilot and now full owner of Venom G-GONE. Dick was a former RAF fighter pilot. They took part in the display season which saw G-BOOM and G-LOSM at North Weald on 16 May, Sabre G-SABR attend the Mildenhall Air Fete on 28-30 May; G-BOOM and G-LOSM fly to the Dutch AF Open Day at Leeuwarden on 11 June; BOOM plus *Heritage Pair* to Cosford on 20 June, Venom G-GONE display at Yeovilton on 16 July (with G-BOOM G-HELV & G-SABR in the static display), flying north to BAe Woodford the following day. G-LOSM flew at Shawbury's Families Day on 29 July, with G-HELV making the regular visit to Lelystad Airshow on 28 August. G-BOOM unusually attended Finningley's Battle of Britain Show on 17-18 September, whilst G-LOSM appeared at Leuchars Show – Jet Heritage were rarely seen at Battle of Britain shows.

Vampire U-1215 was repainted in RAF camouflage marking during the summer of 1993, carrying 215 as identification. The colour scheme was chosen so that the Vampire matched Meteor G-LOSM; they displayed at shows as the *Heritage Pair*. (MHP)

Restoration work on XE677 took longer than anticipated, mainly due to the pressure of other work. Carrying its civil registration G-HHUN by early 1990, the rear fuselage is on its special engineering trolley whilst the front fuselage and wings are to the right. (MHP)

More detailed work being carried out on XE677 later in 1991. The fuselage has been stripped of old paintwork and fully cleaned, enabling the overhauled systems to be re-installed. However, its first flight is still three years away. (JHL collection)

Two further anniversaries were marked in September. 7 September was the fortieth anniversary of Neville Duke capturing the Air Speed Record in Hunter WB188. To mark the event, he flew the course once again at mid-day – this time as a passenger in G-BOOM which was piloted by Adrian – in fact a re-run of 1983. Two fast passes were made over Tangmere where a number of guests had gathered, including Neville's wife, Gwen. Despite having met Neville on a number of occasions, Adrian still remained in awe of him – 'A British gentleman test pilot'. The second event on 20 September was to mark the fiftieth anniversary of the first flight of the de Havilland Vampire. Poor weather put paid to the planned flying display, but Vampire G-HELV was able to put on a short display during the afternoon. The RAF sent some visiting aircraft – Jaguar XX748 from 54 Sqd. (second Vampire squadron) and Hawk XX235 from 4 FTS Valley (a former Vampire station). The event was attended by a number of famous test pilots, including John Cunningham, Neville Duke, Eric Brown, Ron Clear and John Wilson.

Having lost Doug Arnold's collection, Bournemouth Airport gained another one with the arrival of Don Wood's Source Classic Jets on 1 June. Their initial fleet flown comprised three Vampire T.55's, a FB.6 and a Venom FB.50, with a Buccaneer arriving on 19 August. Its planned overhaul for display work was of interest to Jet Heritage. This was because the strike bomber was considered a 'complex' aircraft by the CAA due to the large amount of technical backing it required. However, Jet Heritage's interest came to an end when the CAA advised Source they would never be able to grant the Buccaneer a Permit to Fly due to its complexity. It was advertised for sale in a US magazine of March 1994 but remained unsold.

Financial problems due to recession in Britain and high insurance costs were causing concerns for the directors of Jet Heritage. Due to the high cost of operating vintage jet aircraft, coupled with CAA restrictions, investment was no longer attractive in this type of aircraft, and owners were now seeking to sell them on. Efforts were made to obtain sponsorship for displays, but none was forthcoming,

September 1993 saw Neville Duke and G-BOOM reunited again. To mark the fortieth anniversary of his record breaking flight, Neville returned to Bournemouth in order to be flown over the course once more, this time with Adrian piloting the Hunter. (JHL collection)

so the establishment of a charitable trust was investigated. There was also the directors' aim of establishing a museum in Hangar 600 so the public could see work being undertaken on the aircraft. None of the active aircraft were owned by Jet Heritage – G-HELV and G-LOSM were both registered to Hunter Wing, G-GONE to John Davies and G-SABR to Golden Apple. Ian and Douglas had already sold their Jet Provosts and negotiations were in hand to sell G-BOOM. Purchasers were sought for Jet Heritage from July, where it was hoped that it would continue as a charitable museum. Consideration was also given to moving to another airfield such as Kemble or North Weald where hangarage would be less expensive. Interest was shown by Short Bros in leasing Hangar 600, but this was not an ideal situation. It was back in December 1992 that the directors had commenced an application to form a charity which would have financial benefits in connection with a proposed museum. The charity would assist in the fund-raising aspects of the operation and be able to apply for grants – something the company could not do. At long last the application was agreed, seeing the formation of the Jet Heritage Charitable Foundation at the beginning of 1994. Its main aim was to preserve and present to the public operational examples of Britain's vintage jet aircraft. Further aims were to seek passionate aviators to sit on the board of trustees, to build a museum for a national collection of operational historic jet aircraft and to obtain funding for the museum.

G-BOOM and G-HELV undertook some circuit flying at Bournemouth on 25 November 1993, prior to another Sotheby's Auction at Billingshurst on 27 November. Again a number of the Jet Heritage aircraft were entered, but failed to reach their reserve prices. The bids were: unflown Hunter G-HHUN £55,000; Provost XF877 £24,000 and Gnat G-NAAT £4,200. The exception was Hunter G-BOOM which was sold to Richard Verrall of RV Aviation for £130,000 (a value of £200,000 had been quoted in July). The reason for the high price was that the Hunter was needed for a new project.

Historic aircraft visiting Bournemouth Airport would often stop off at Jet Heritage's apron. Here BoBMF Hurricane PZ865 is parked outside Hangar 332, attracting attention from the engineers and nostalgia for Eric – a former Hawker man. (J Atkins)

12

The Royal Jordanian Air Force and 'Harry'

King Hussein of Jordan was an experienced pilot and keen supporter of his country's air force. In the early 1990s the king was promoting the establishment of a Royal Jordanian Air Force museum, mentioning this during his visits to Great Britain. Richard Verrall was a civil pilot fully employed by British Airways. He was also an experienced helicopter pilot, being loaned out by British Airways as pilot of the king's Agusta helicopter and other aircraft whenever he was visiting Britain. Richard had also piloted Brencham's Agusta A.109 from Biggin Hill back in the 1980s. During a conversation with the king, Richard learnt that he was keen to establish the museum, which would include an Historic Flight attached to the Air Force. The Flight was 'to reflect the Royal Jordanian Air Force's distinguished heritage of the past four decades'. Richard had his own consultancy company, RV Aviation, which in 1993 was contracted by the Air Force to form the Flight which would eventually be based in Amman. Jordan had received Vampire fighters and trainers in 1955 and Hunters in 1958, so the king wanted these types to be included in the Flight. RV Aviation set about seeking some out – it seeming appropriate that Britain was again supplying fighters to Jordan, as it had back in the 1950s. RV Aviation's original plan was to set up an overhaul facility at North Weald, but Richard saw the potential connection with Jet Heritage. Following the purchase of G-BOOM in November 1993 as the Flight's first aircraft, it was agreed that overhaul work would be undertaken by Jet Heritage. It was seen as a wonderful opportunity to promote the company, as well as a potential financial lifeline. The driving force behind the Historic Flight was King Hussein, and he visited Bournemouth with Richard on 14 December 1993 to see the work currently being undertaken by Jet Heritage.

Eric: The king arrived by helicopter in the early afternoon and was greeted by the Jet Heritage VIPs. I was rather in awe, but being Chief Engineer was elected to escort him around the hangars. At first I felt out of my depth, but soon found him to be a warm and well advised enthusiast who knew what he was talking about. Progressively, our relationship changed to friendship, although I was constantly taken aback when he called me 'sir' – presumably because of my age! The visit ended in the main reception office from where our secretary Carol James decided to beat a hasty retreat. The king

1 Spencer Flack's determination saw Hawker Hunter G-HUNT take to the air from Elstree on 20 March 1980. (EGH)

2 Fully painted in Spencer's house colours, G-HUNT on an early air show outing to Alconbury in August 1980. (JHL collection)

3 Stephan Karwowski departing Biggin Hill in G-HUNT, May 1980, no doubt followed by a dive into Biggin's adjacent valley. (JHL collection)

4 The second Hunter restoration undertaken during 1980 was Brian Kay's G-BOOM, seen at Stansted the following summer. (JHL collection)

5a Cover of pamphlet produced by Hunter One for the 1981 display season.

5b The initial Hunter One team. G-HUNT and G-BOOM (still in original colours) along with Adrian Gjertsen, Eric Hayward and Mike Carlton. (JHL collection)

6 Now in Hunter One ownership and painted red overall, G-HUNT overflys the Needles in a classic Arthur Gibson photo. (A Gibson)

7 The 1983 Hunter One Display Team: Eric Hayward, Geoff Roberts, Mike Carlton, Adrian Gjertsen and Brian Henwood. (JHL collection)

8 In order to match G-HUNT, Mike Carlton had Hunter Trainer G-BOOM repainted red for the 1983 display season. (MHP)

9 Neville Duke's first speed record re-run in September 1983. Mike Carlton chats to Neville, with his wife Gwen alongside Adrian. (A Gjertsen collection)

10 G-HUNT on one of its visits to BAe Dunsfold's Families Day. Mike and Adrian with retired Hawker test pilot David Lockspeicer. (EGH)

11 Two red Hunters – but displayed as Hunter One. G-HUNT and G-BOOM flying over Dorset's Purbeck Hills.

(A Gibson)

12 G-HUNT at the Bournemouth Air Show in July 1983, while its ground crew watch the Vulcan display. (JHL collection)

13 Flagship of Hunter One, G-HUNT, with its overall red colour scheme seen to full effect whilst on a visit to Coventry. (JHL collection)

14 Gloster Meteor WM167 as viewed by Hunter One at Blackbushe in the spring of 1984, still in its target towing configuration. (MHP)

15 Repainted into its former RAF camouflage scheme, WM167/G-LOSM now wears the markings of 141 Squadron. (JHL collection)

16 Jet Provost 352 was one of a number marked for disposal by the Singapore Air Force in 1984. Along with 355 it was purchased by Hunter One. (EGH)

17 G-JETP was the second of the former Singapore Jet Provosts to be overhauled by Hunter One, flying again in July 1985. (Peter R March)

18 Three of the Hunter One fleet over the Dorset countryside: Hunter G-BOOM, Jet Provost G-PROV and Meteor G-LOSM. (Peter R March)

19 & 20 Jet Provost G-PROV (above) over Dorset and G-JETP (below) over Poole Harbour on a photo shoot in the summer of 1985. (Peter R March)

21 Hunter Trainer G-BOOM in its washable paint scheme for filming *Return to Fairborough* in October 1984. (MHP)

22 Wilf Hardy's painting for 1986 TVS Air Show depicting G-PROV leading G-AWPH, G-HUNT and G-VIXN – something that did not happen at the show. (RAF Benevolent Fund)

23 Sold in October 1987, Hunter G-HUNT ended up in the safe hands of Jim Robinson at Houston, Texas. (JHL collection)

24 Jet Heritage was launched at Bournemouth in February 1989 to carry on the pioneering work of Hunter One. (Allan Burney)

Above: 25 By the time it was operated by Jet Heritage, Meteor G-LOSM had lost its 141 Squadron markings. (Peter R March)

Right: 26 DH Venom G-GONE was hangared with Jet Heritage by Philip Meeson. It wears Royal Navy 'Admiral's Barge' colours. (JHL collection)

27 Percival Provost G-AWVF was intended to be displayed along with Jet Heritage's Jet Provost, but the plans did not materialise. (JHL collection)

28 The 50 Years of British Jet Flight tribute at Bournemouth in May 1991 saw a gathering of Britain's greatest test pilots. (A Gjertsen collection)

29 North American F-86 G-SABR was maintained by Jet Heritage on behalf of the Golden Apple Trust. (JHL collection)

30 Following its arrival in England, Adrian Gjertsen soon acquired a soft spot for flying the F-86 Sabre. (A Gjertsen collection)

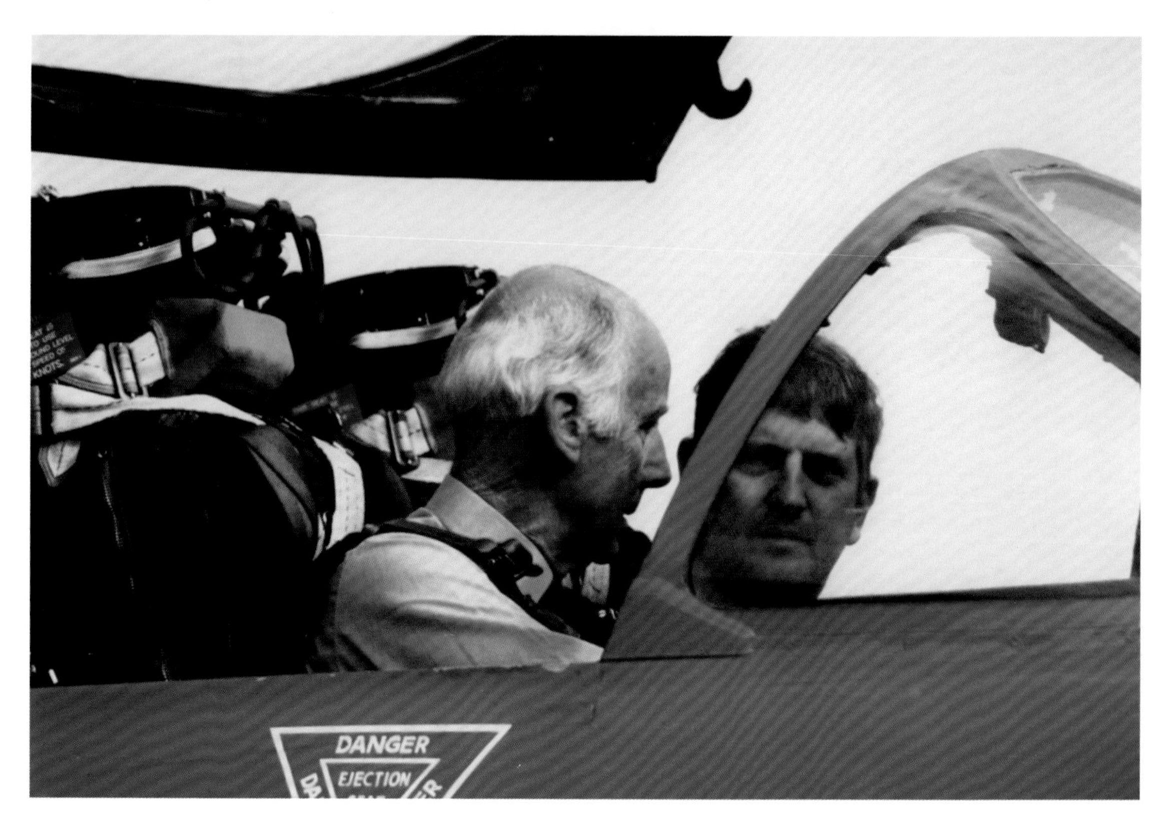

31 Neville Duke's second speed record re-run, this time in September 1993 with Adrian piloting G-BOOM. (JHL collection)

32 Hawker Sea Hawk WV795 was held by Jet Heritage as a possible long-term restoration project. (MHP)

33 Destined for the Jordanian Historic Flight, Vampire J-1106 arrives at Bournemouth in its bright target-facility colour scheme. (Peter R March)

34 Jet Heritage engineers at work on Swiss Vampire J-1106 during the summer of 1994. (JHL collection)

35 New flagship Hunter XE677 flew again in January 1994 in a colour scheme similar to Neville Duke's historic WB188. (Brian S Strickland)

36 Folland Gnat XR537 in its Red Arrows colour scheme. Its overhaul proved to be more protracted than anticipated. (MHP)

37 & 38: The Swiss are here! Hunters J-4208 and J-4081 arrive at Bournemouth on 16 June 1995, led by G-BOOM/800 (above). J-4075 and J-4081 follow with XE677 (below). (Peter R March)

39 Busy scene on the Jet Heritage apron following the arrival of the Swiss Hunters in June 1995. (Peter R March)

40 The Swiss Air Force pilots who delivered the Hunters pose with Brian Henwood in front of J-4208. (Brian S Strickland)

41 Wearing matching camouflage colour schemes Meteor G-LOSM and Vampire G-HELV were promoted as the *Heritage Pair.* (JHL collection)

42 Overhaul underway on one of the Swiss Hunters - Bob Tarrant checking pipe work behind the wing leading edge. (JHL collection)

43 Hunter XE677 at Guernsey Airport in September 1998 for the Channel Islands' Battle of Britain display. (JHL collection)

44 The Jet Heritage hangar complex at Bournemouth in 1994 with the majority of the fleet parked outside. (JHL collection)

45 *Heritage Pair* Vampire G-HELV in RAF markings carrying 215 as its identification number. (JHL collection)

46 Jet Heritage engineers prepare Hunters XE677 and G-VETA for flight on a bright but chilly winter's morning. (MHP)

47 Hunter Trainer G-HVIP inside Hangar 600 in the spring of 1996 prior to hand-over to its German owner. (MHP)

48 Hunter 800 and Vampires 109 and 209 at Bournemouth, prior to their delivery to the Royal Jordanian Air Force Historic Flight. (Peter R March)

49 Jordanian Vampire Trainer 209/G-BVLM crew training at Bournemouth, spring 1997. (A Harris)

50 Jordanian Hunter 843/G-BWKC on the flightline carrying the four underwing fuel tanks fitted for its delivery flight to Amman. (MHP)

51 Hunter 800/G-BOOM crew training Jordanian pilots at Bournemouth in the spring of 1997. (A Harris)

52 The Jordanian Royal Flight VIP Dove was a frequent visitor to Jet Heritage during the Historic Flight contract. (EGH)

53 The opening of the Jet Heritage Museum in May 1998 coincided with a visit by Concorde to Bournemouth. (MHP)

54 A trio of Hunters on the flight line after their display to mark the Jet Heritage Museum opening. (MHP)

55 Jordanian Hunters 800 and 843 prepare to depart Bournemouth in May 1998 on delivery to Amman. (MHP)

56 Hunter Trainer WV372 after restoration to its RAF colour scheme of the 1960s. The aircraft was operated by a syndicate of pilots. (MHP)

57 Hunter G-VETA was restored by Jet Heritage on behalf of a syndicate of Cathay Pacific pilots. (MHP)

58 Jonathon Whaley's very colourful Hunter G-PSST *Miss Demeanour* at the Jet Heritage Museum Open Day on June 1999. (MHP)

Above: Hunter F.6's 704 and 710 in service with 1 Sqd, Royal Jordanian AF during the 1960s. The colourful fin and wing tip markings were not carried all the time by 1 Sqd, so the camouflage markings applied by Jet Heritage to the Historic Flight Hunters were accurate. (RJAF)

Right: In connection with the Jordanian Historic Flight contract, King Hussein first visited Jet Heritage in December 1993. He was most impressed with the engineering work being undertaken and with the future proposals for the Flight, happily signing the visitors' book. (EGH)

asked to sign the Jet Heritage Visitors' Book, noting he was, 'most impressed on my memorable visit'. Further meetings and visits were to follow.

To increase the planned Historic Flight fleet, further Hunters and Vampires were sought. Eric and Richard visited Switzerland early in January 1994 to view some Vampires at Altenrhein, with a further visit at the end of March to check on a Vampire T.55. Eric and Richard then made a visit to Jordan at the end of February to see what the set-up in Amman was like, being guided by Brigadier General Nabil Barto. They were flown around the bases in an Air Force Blackhawk helicopter.

Eric: I had been asked to look over a Hunter stored at Mafraq, as it was considered suitable to join the Historic Flight. It seemed in reasonable condition, and I recommended that it be dismantled and flown by Hercules to Bournemouth for a full survey. It proved to be RJAF 842, having recently been presented to Jordan by the Omani Air Force. Then it was on to Prince Hassan Air Base, where there were six F-104 Starfighters. Four were in quite good condition, and I suggested that two could be used as static exhibits in a future RJAF Museum. Then to Marka to examine the reported store of Hunter and Vampire parts. This proved to be a dead end, as there was hardly anything. In the past the Omani Air Force had helped Jordan with the supply of aircraft and parts, and I suggested that further support could be sought by way of Hunter spares. The following day I met King Hussein again to discuss various matters. He agreed to the Hunter being returned to Jet Heritage and also to request Oman for assistance. This included parts and an engine for the possible upgrade of 800/BOOM. In the end it turned out that Oman had no spares available. Additionally, further examination of Hunter 842 revealed that it had a heavy sand coating after years of storage, and so it was decided not to restore the airframe.

G-BOOM was one of the Hunter trainer variants fitted with the 'small' Avon 100 engine, whereas those that served in Jordan were later ones with the higher-powered Avon 200 engine. Consideration was given to rebuilding G-BOOM to take the 200 engine, but it was eventually agreed that it would be too complex a task within the given budget. G-BOOM was repainted in RJAF colours during May 1994 with serial 800 and the markings of 1 Squadron. The second aircraft for the Flight was former Swiss AF Vampire T.55 U-1216/ZH563, which had been owned by the RAF Benevolent Fund but was acquired by RV Aviation in April. I had been originally inspected by Jet Heritage back in January 1992, and a team was dispatched to Boscombe Down to prepare it for a ferry flight to Bournemouth on 19 April. At the time it was unique in carrying three different identities at the same time – U-1216 (Swiss AF), ZH563 (British military) and G-BVLM (British civil). By July it had been repainted in camouflage colours as RJAF 209 in the markings of 2 Squadron. In connection with the work being undertaken at Jet Heritage, there were frequent visits by Jordanian Royal Flight DH Dove 8 JY-RJU and Sikorsky S-76 G-BTLA.

Above: Following his visit to Bournemouth, King Hussein invited Eric and Richard Verrall to Jordan to tour military bases to see what was available for the Historic Flight. Hosted by the king, further planning of the Flight took place in his palace. (EGH collection)

Right: During their tour of various air bases around Amman, Eric and Richard were accompanied by Brigadier General Nabil Barto. He is seen with Richard outside the headquarters of the Jordanian Royal Squadron. (EGH collection)

The first aircraft for the Jordanian Historic Flight was Hunter G-BOOM. This was repainted from its well-known red colour scheme into Jordanian Air Force camouflage as 800 during May 1994, soon to be joined by a similarly painted Vampire. (MHP)

Vampire Trainer G-BVLM was to be the second Historic Flight aircraft. U-1216 with the Swiss AF, it was donated to RAF Benevolent Fund and became ZH563. It then passed to the Historic Flight with civil registration G-BVLM, all three markings still being carried in April 1994. (MHP)

After overhaul, Vampire G-BVLM was repainted in Jordanian colours as 209. Along with Hunter 800, the Vampire appeared at British air displays during 1994 to promote the newly formed Jordanian Historic Flight. (MHP)

The summer of 1994 saw both G-BOOM/800 and G-BVLM/209 available for the British air show season. The second Vampire for the Flight was FB.6 J-1106/HB-RVO, which flew into Bournemouth from Altenrhein on 30 June 1994. At the time its fuselage and wings were painted in a vivid black and orange day-glo colour scheme, as it had been used by the Swiss as a target presentation aircraft. Re-registered as G-BVPO, its overhaul was under way by the time the Chief of the Royal Jordanian Air Force visited Jet Heritage on 28 July. Work included removing the pointed 'Swiss' nose and replacing it with the original 'de Havilland' version. The chief was also given an experience flight by Brian Henwood in Hunter 800. Eric undertook further trips in September to view Hunters – swift visits to Berne and Sion, followed by Scampton in October. Here Richard Verrall had discovered maintenance airframe XG160/8831M in good condition, and wanted Eric's opinion. Following his favourable report, the Hunter was presented by the RAF to Prince Feisal in December for the Historic Flight, and moved to Bournemouth by road on 21 January 1995. Now painted matt black in 111 Squadron colours, it was part of their 1958 twenty-two aircraft formation loop piloted by Sqd. Ldr Roger Topp. Unfortunately, it soon had this paint scheme stripped off prior to overhaul, with registration G-BWAF allocated. Vampire G-BVPO was flown again by Dick Hadlow on the afternoon of 22 December, painted as RJAF 109 in the camouflage colours of 2 Squadron.

Jet Heritage engineers attending to the other Vampire trainer in their fleet, former Swiss Air Force G-HELV/215. Along with Meteor Night Fighter G-LOSM it formed one half of the *Heritage Pair.* (JHL Collection)

Jet Heritage's workload suddenly increased with the arrival of four Swiss Air Force Hunters in June 1995. Here the formation arrives at Bournemouth, led by 800/G-BOOM with XE677 bringing up the rear. Even a de Havilland man might have sighed 'Hah, Hawker!' (Peter R March)

A scene of great activity on the Jet Heritage ramp following the arrival of the Swiss Hunters. Naturally, there was much interest in this activity, with a number of visitors arriving to inspect the aircraft over the next few weeks. (Brian S Strickland)

King Hussein and his son, Prince Hamzah, inspecting the cockpit of Vampire J-1106, which was destined for the Historic Flight. Security had to be tight for such occasions, and it appears to have worked, as news of these visits did not slip out to the media. (EGH collection)

One of the reasons for Eric's visits to Switzerland was to survey various Hunters which had been withdrawn from service by the Swiss Air Force in December 1994 and were to be disposed of in the spring of 1995. The arrangements were different from the previous Vampire sales. Interested museums and societies were invited to write in their reasons for wanting a Hunter. The requests were considered by a committee who decided the allocation of the aircraft by way of gifts. The Air Force would then deliver each aircraft, fuelled for one flight only, to an airfield of the recipients' choice. The pilot and aircraft's VHF radio equipment had to be returned to Switzerland at the recipients' expense. Any ongoing flights were also the new owners' responsibility.

Eric realised that this was another opportunity for Jet Heritage's expertise to be used. A quick phone call to his air force contacts in Emmen resulted in a further visit to Switzerland in May. Here Eric selected four Hunters – two F.58s for Jordan, one for Jet Heritage and a T.68 for a private owner – and arrangements were made for their ferry flight. Flown by Swiss Air Force pilots, they arrived in England en masse on 16 June 1995. J-4075/81/83 and J-4208 were met on arrival over the coast at the Mayfield beacon by G-BOOM/800 and XE677, put up by Jet Heritage and flown by Brian Henwood and Adrian. The formation then overflew Dunsfold airfield as a salute to their place of birth before arriving mid-day at Bournemouth. Then it was smoke on for a run and break prior to a stream landing. The airport hadn't seen a six-ship flypast down the runway for some while! A fifth Swiss Hunter, J-4025, was flown to Fairford having been presented to the RAF Benevolent Fund. It was on static display at the RIAT Display at the end of July, with ownership passing to the Jordanian project, and flown down to Bournemouth on 24 July. The Hunters were then parked to await their turn for restoration.

Now to the Harry of the title. After his first visit in December 1993, it became apparent that King Hussein was going to take a keen interest in progress. This posed various security problems, as his visits needed to be low key. So, with the agreement of his security staff, Jet Heritage referred to him as 'Harry'. Therefore, any notes and memos did not indicate reference to the king.

27 May 1994 saw a surprise visit in the late afternoon by the king and his wife, Queen Noor, with Richard, as ever, having phoned in advance to warn of a visit by 'Harry'. On 31 May King Hussein returned again with his son Prince Hamzah to view the aircraft, the young prince enjoying sitting in the cockpits, including that of the newly repainted G-BOOM/800. In August King Hussein wrote to Eric saying he would be delighted and honoured to become the Royal Patron of the newly formed Jet Heritage Charitable Foundation. On 15 February 1995 his son visited Jet Heritage again in company with Richard Verrall to see progress. This was followed on 11 May by King Hussein in company with Major General Ababneh and Brigadier General Nabil Barto of the Royal Jordanian Air Force. They were all pleased with what they saw.

Wishing to build on their royal patronage, Jet Heritage laid on a VIP luncheon in Hangar 600 on 6 July 1995 hosted by King Hussein. The menu included a choice of 'Rolls à la Meteor, Fillet de Bœuf Vampire, Salade de Riz au Hunter, ending with Afterburner (sur demande)'. The guests were people known to be interested in historic aircraft, and the intention was to encourage them to donate funds to the charitable foundation. The king promoted the organisation, saying that Jet Heritage had a great future. He ended by offering the initial donation to the charity. Unfortunately, none of the VIPs did likewise at the table, nor were they requested to do so during the afternoon's events. So no funding resulted from the luncheon which would have ensured the future of the Jet Heritage Charitable Foundation. At the time the launch of National Lottery funding was still in the pipeline, and was not a route that could be explored.

Among his many aviation interests, King Hussein was patron of the Royal International Air Tattoo which was held at Fairford. Later in July, while he was in the country, he attended RIAT, where Jordanian Hunter 800 was displayed along with the two Vampires. A special trip was laid on for the king to visit the Jersey Air Show on 14 September, flying out, with others, in his Dove JY-RJU. On the way home a night stop was made in Guernsey, where discussions were held with Frank Hackett-Jones regarding hangaring the two RJAFHF Vampires for the winter months, as Hangar 600 at Bournemouth was full.

Richard and Eric flew to Cranwell in Dove JY-RJU on 15 June 1995 in order to view surplus RAF Hunters. 22 July saw a visit to Bournemouth by Duncan Simpson, a former Hawker Siddeley Test Pilot, accompanied by Prince Feisal. In October three of the former Swiss aircraft were registered to RV Aviation as G-BWKA-KC with the fourth as G-EGHH (personalised registration for Eric G Hayward's Hunter) in July 1995. As an indication of prices, Vampire G-BVPO was valued at £66,000 in July 1994 and Hunter G-BWAF at £100,000 in February 1995.

The two Vampires returned from winter storage in Guernsey on 30 March 1996, along with Dove JY-RJU. In May a group of Jordanian Air Force engineers arrived for a three-week familiarisation programme on Hunters and Vampires. Three of them had knowledge of Hunters, but none of them had worked on Vampires. The first of the former Swiss aircraft, G-BWKA, was completed in May, emerging as RJAF 843 and flying on 13 June. This was followed by G-BWKC, which flew as 712 in August – both wearing the markings of 1 Squadron. A visit to Airwork Service Training at Perth was made in June to survey an instructional Hunter, followed by RAF Swinderby in October to view Hunter XL577. In the event, neither aircraft were suitable for the Jordanian contract. Hunter 800 and Vampire 109 both flew at Woodford's RAFA show on 24 June, with Vampire 209 and the Dove on static display. Then 800 and 109 appeared at Waddington's show on 1 July. At RIAT Fairford on 21 July 1996, there was a mass flying presence of the Jordanian aircraft in front of King Hussein – Hunter 843 and Vampire 109 on the Saturday, joined by 800 and 209 on the Sunday. The Fairford show was followed by a further visit to Hangar 600 by King Hussein on afternoon of 26 July to see progress on the aircraft. On another of his visits he spotted a Second World War jeep that Eric had recently acquired. The King was so impressed by Eric's demonstration drive around the Jet Heritage apron that he later purchased one for Queen Noor.

Above: Former Swiss Hunter J-4075, now resplendent in Royal Jordanian Air Force colours as 843. During 1996, the Historic Flight Hunters and Vampire visited a number of British air shows, prior to being delivered to Amman the following year. (MHP)

Right: King Hussein was an aircraft enthusiast and he got on very well with Eric Hayward – both men fully understanding what they were talking about aviation-wise. They are seen with Eric's Jeep which the king took a fancy to, eventually buying one for his wife. (EGH collection)

13

Organisation Changes

The setting up of the Jet Heritage Charitable Foundation early in 1994 would hopefully have put the future of Jet Heritage on a firmer footing. Constitutionally the foundation comprised Jet Heritage Ltd (which maintained and operated the aircraft) and the proposed Jet Heritage Museum. Its aim was, 'the advancement of public education in the jet heritage of the United Kingdom of Great Britain & Northern Ireland by the restoration, preservation and exhibition of historic jet aircraft'. Such grand-sounding aims always appealed in applications made for Lottery Funding. It was planned to keep at least one example of a Meteor, Vampire and Hunter flying for as long as possible, to flight-display the aircraft at air displays and at Bournemouth. A static display of aircraft and engines was to be established for the museum, with training of future pilots and engineers undertaken on the fleet. If sufficient funding could be raised, a further aim was to purchase the freehold of Hangar 600, plus the adjacent Hangar 332, so giving the foundation a more secure future.

After months of restoration, the afternoon of 21 January 1994 saw Adrian undertake the first flight of Hunter XE677/G-HHUN. At the time still in bare metal, apart from its underwing tanks, this was its first flight for thirty-three years. Further flights were undertaken before the aircraft was painted red in early April (thereby replicating WB188 and G-HUNT) with one of its first displays at Middle Wallop on 15 May. Other air display work saw XE677 and G-LOSM appearing at the Biggin Hill Show on 18-19 June. A photo shoot was then set up on 8 July on behalf of John Dibbs. He arrived from Duxford in Grumman Avenger G-BTDP, which was his mount to photograph Hunters XE677, G-BOOM/800 and Vampire G-BVLM/209. The *Heritage Pair* displayed at Farnborough Air Show in September. Not to be outdone, FRA from Bournemouth sent a four-ship Falcon 20 formation!

The F-86 Sabre G-SABR moved to a new operating base with the OFMC at Duxford in March 1994, the new site being more convenient for its owner. Another departure was Venom G-GONE, which flew out in November to Hawarden for a period of storage on behalf of John Davies. Arriving by road from Shawbury just before Christmas was Hunter GA.11 XF301, which had been purchased by an American collector. He required some work on the airframe, including the addition of a tail-braking parachute, before it was shipped to the States.

Work underway in Jet Heritage's engineering hangar in the summer of 1994. At the time the aim was to get the Jordanian aircraft flying and ready for hand-over. This meant that work on the Swift (background) and Sea Hawk (right) was suspended. (JHL collection)

The overhaul of Hunter XE677 was completed at the beginning of 1994. Here the unpainted aircraft, registered G-HHUN, is being fuelled prior to its first flight piloted by Adrian. Its last RAF flight was back in 1961, when it flew into Dunsfold. (JHL collection)

Following its test-flights, XE677 was immaculately painted red overall – a colour selected by Eric. As such it resembled Neville Duke's record-breaking Hunter prototype and, rightly, it turned heads whenever it appeared at a display. (MHP)

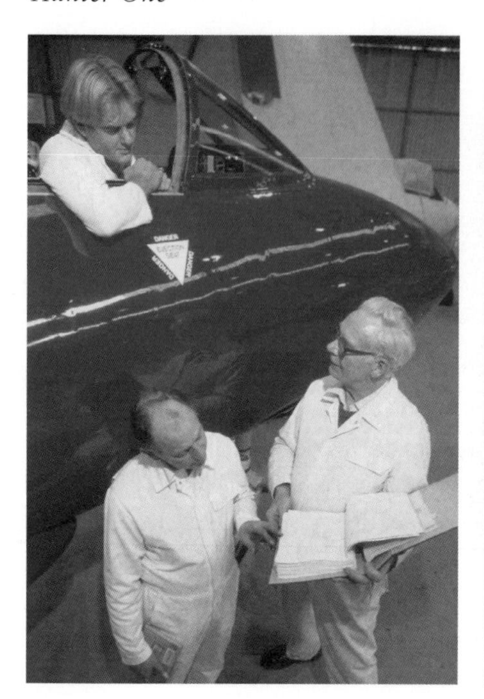

(Left) Inside the hangars, Eric is going through details in a Hunter AP Manual with two of the engineers who are working on XE677. (Right) The stored Sea Hawk is seen with spare Vampire cockpits in the foreground. (JHL collection)

With all the additional engineering work being undertaken, Eric was pleased to receive some experienced help in the shape of Bob Tarrant in January 1995. At one time with British Aerospace at Bournemouth, Bob was well experienced on Hunters having arrived from FRADU, Yeovilton. In March a message was received to say that a German doctor was going to visit from Stuttgart. This turned out to be Karl Theurer, who arrived in his Mu-2 D-IKKY on 20 March. However, this was not a medical visit – he was about to buy a Hunter. An experienced pilot, he was aware that the Swiss Air Force was disposing of its fleet, but he was hampered by German regulations, which prevented former military jets being owned or operated by civilians in Germany. So could Jet Heritage help him by restoring one and then housing it for him? Negotiations led to T.68 J-4208 being acquired by Karl, but it needed to be ferried to England. When Eric visited Switzerland in May he arranged that Karl's aircraft would be flown across, along with the three prospective Jordanian aircraft. So, J-4208 arrived on 16 June as part of the mass formation, being registered as G-HVIP (representing Hunter VIP) in July to Golden Europe Jet De Luxe Club Ltd., a Jersey-registered company. Even so, this still did not enable Karl to fly a Hunter in Germany, added to which, he was not qualified on jet fighters. So, about once a month he would fly to Bournemouth in his Mu-2 to obtain conversion training on G-BOOM/800.

Work on its overhaul having commenced back in 1990, another effort was made in the spring of 1995 to get Gnat XR537 back into the air. Although it appeared ready to fly, a number of snags, including a non-functional ejector seat, kept it grounded. (MHP)

The spring saw Gnat XR537 being dusted off, and further efforts were made to get it airworthy on behalf of Frank Hackett-Jones. He now considered undertaking a ferry flight from Bournemouth to his Guernsey base for further restoration. Lack of a working ejector seat brought an end to this plan. By June Frank was considering a low-level flight only, but Eric suggested that it would be better to ship it by ferry from Portsmouth. However, XR537 remained with Jet Heritage, with consideration given to installing an electrical start system, as opposed to using the normal Palouste GPU. There was another overseas trip for Eric in September 1995, when a team from Jet Heritage flew to Guernsey to attend a party given by Frank for his aviation contacts and friends.

From the spring of 1995 the Chipmunks of the RAF's No.2 Air Experience Flight were housed in Hangar 332 by Shorts, also sharing the Jet Heritage ramp until March 1996. Philip Meeson's Dragon Rapide G-AGSH had been away on overhaul for some time. However, it returned on 10 June, resplendent in BEA colours, receiving a fresh C of A on 20 June. Also in June, Jet Heritage were proud to announce that they had received CAA overhaul A8-20 approval following the high standard of past maintenance work on the fleet.

Adrian: For some months I had not been happy with changes that were occurring within Jet Heritage. I felt that it was now heading in a different direction from that

Meteor F.8 VZ467 was inspected at Cosford in spring 1994, prior to being flown to Biggin Hill by Adrian Gjertsen. Now registered G-METE and operated by Adrian's Classic Jets, it is seen inside the former Brencham hangar shortly after its arrival at Biggin Hill. (MHP)

> I had envisaged back in 1988. This is bound to happen within any company where the other shareholders had to look after the needs of the financial investors. I also felt there was less opportunity for me to undertake air display work. So I set up a new organisation – Classic Jets – based at Biggin Hill and initially equipped it with Meteor F.8 VZ467/G-METE and Strikemaster G-SARK. Matters came to a head at Jet Heritage in July 1995 and I officially resigned as a director on 31 July. My place was taken by Ian and Dougal.

Anticipating his move, Adrian had purchased Meteor F.8 VZ467 which had been in storage at Scampton. The aircraft was crated and transported by road to Cosford, where it underwent an RAF major servicing over a couple of years, under the watchful eye of Chief Tech. Merv Roberts. Eric then checked it over in May 1994 prior to being able to obtain the necessary CAA permit for a ferry flight. Initially this was reported as being to Bournemouth, but in the event the Meteor was flown to Biggin Hill to be the first aircraft of Classic Jets. As such it was displayed at the Air Fair on 18 June. Adrian had already realised that he would not be able to complete the restoration of his Sea Hawk G-SEAH. He was in contact with Wally Fisk, an American warbird dealer who owned Amjet in Minnesota. Wally was interested in naval aircraft and Adrian arranged a deal with him. In exchange for his Sea Hawk, Wally would supply him with an airworthy BAC Strikemaster. G-SEAH was packed into crates and departed by sea in November 1994, although the Strikemaster did not arrive until the following year. As had been the case with Adrian's Sea Hawk, both new aircraft were registered to Sark International Airways.

Meteor G-LOSM made an early Continental show appearance at Eindhoven on 30 April, then displaying at Gosselies on 21 May along with G-HHUN. G-HELV was on show at Chièvres on 25 June, where it was joined by four of Source's Vampires.

Back home, XE677 displayed at Shawbury on 27 July 1995, G-PROV at North Weald on 5 September and G-LOSM at the Biggin Hill Air Fair on 17 September. The rather dilapidated Sea Hawk WV795 and Whirlwind XN385 which had been in storage for the Cypriot Museum finally departed – but not for Cyprus. Their owner appeared to have lost interest in his aircraft and they departed for Bruntingthorpe in August. He had already bought two ex-RAF Shackleton AEW.2s which were flown to Cyprus where, unattended, they deteriorated over the following years.

Following the disposal by the Swiss Air Force of their Hunters in 1995, a number were destined for the United States. As part of the arrangement, the Swiss agreed to deliver them as far as Southend in July, with the various American owners arranging the onward flight to the States. About eight arrived, but there were problems with the aircraft which required Jet Heritage's experience.

Eric: Following the arrival of the aircraft at Southend, the local police somehow got to hear that the Hunters still had their Aden cannons fitted. This was true, although they had been disabled by the Swiss Air Force. This was not good enough for the police, who impounded the aircraft within a fence and demanded that their guns be removed a.s.a.p. Nobody at Southend was able to undertake the work, so it fell to Jet Heritage to solve the problem. So in October I sent a team to Southend to see what was needed. It was agreed that four of the Hunters would have their guns removed, ballast added and then flown to Bournemouth for further attention. The work was duly undertaken and the team returned to Bournemouth. Having seen the work being carried out, the police lost interest in the guns. Perhaps they should have paid more attention as they returned to Bournemouth in the back of our van! The new owners of a couple of the other Hunters got fed up with the problems and sold their new acquisitions to British owners.

J-4060/NX58WJ (California owner) and J-4097/NX58HH (Texas) were ferried to Bournemouth on 27 October, followed by J-4035/N159AM (California) on 1 November with J-4061/N58MX (Pennsylvania) following in March. Its delay was due to the fact that the delivery pilot had damaged the rear fuselage on landing at Southend. The aircraft was not fitted with its gun pack ballast which threw out its centre of gravity, so it required repairs before flying again. At Bournemouth the correct amount of ballast was added by Jet Heritage to replace the guns on the other aircraft, thereby enabling the Hunters to make their way across the Atlantic. At the time, space in the hangars was still proving rather cramped, and it was necessary to store some of the Jet Heritage fleet in other hangars. With the American aircraft out of the way, work was able to commence in earnest on the overhaul and conversion of J-4208/G-HVIP for Karl.

Eric: One of the ferry pilots went by the name of 'Sharkbait Delaware'. He departed with NX58HH on 13 November en route to Texas via Reykjavik, Goose Bay and Labrador. Whenever a ferry pilot departed I gave them my personal telephone number in case a problem occurred in the early stages of the flight. Sure enough,

During the autumn and winter of 1995, further former Swiss Air Force Hunters arrived at Jet Heritage. Destined for the USA, they needed their Aden guns removed before being allowed to continue their journey. J-4060's new registration of NX58WJ is carried in minute letters. (MHP)

I received a call from 'Sharkbait' the following day – the conversation going something like this:

SB “Hi, I'm at Goose Bay. No problems so far but now we can't start the aircraft. When I push the starter button nothing happens.”

EGH “Is it cold out there?”

SB “About minus 20°C!”

EGH “What was the temperature at 36,000 feet during the ferry flight?”

SB “About minus 25°C.”

EGH “Have you got a warm air blower?”

SB “Sure have. Had it in the cockpit to warm me.”

EGH “Don't have it in the cockpit. Open the starter bay door and put the blower in there for ten minutes and then try again. Call me back”

SB *[Later]* “Gee, that did it! Hey, what did I do?”

EGH “Your flight at 36,000 feet gave the systems a long cold soak – probably on landing still well below freezing. Moisture in the IPN fuel froze until the system was warmed up. Result – no start.”

SB “Thanks. Will be away shortly. Will call you again if I have any more problems.”

Hunter Trainer G-HVIP was one of those that arrived from Switzerland in June 1995. After restoration by Jet Heritage for its owner, it received a magnificent blue and white colour scheme, flying again the following April. The registration represented H(unter)VIP. (MHP)

Hunter XL600 on re-assembly at Bournemouth, following its arrival from Fleetlands in June 1996. Tree felling is underway in the background as the first stage of preparing the ground for what was to become a public viewing area. (JHL collection)

He never phoned back and subsequently told me that the rest of the flight went according to plan.

All part of the service that Jet Heritage provided! I later learnt that 'Sharkbait' safely arrived in Arlington two days later.

In January 1996 Eric told Frank Hackett-Jones that hopefully his Gnat XR537 should be flying within a few months. But there were still problems in obtaining spares, including an ejector seat and satisfactory fuel hoses. This showed up the problem of working on other people's aircraft, where they were working to a budget. Frank had a figure in mind which had already been passed. As no fixed price could be quoted for finalising the restoration, Frank regrettably decided to call it a day and work was suspended once again. By the beginning of February 1996 Hunter XF301 had been packed in its crate and dispatched across the Atlantic to Texas for its owner. The last of the American Hunters – J-4061/N58MX - finally departed on 15 April with Brian Grant as pilot for the long delivery flight to Pennsylvania. Karl's Hunter Trainer G-HVIP first flew in April, painted in a light blue and white colour scheme. This resulted in more frequent visits by Karl from Stuttgart to fly his new 'toy' – the first occasion being 30 April. On 4 May G-BOOM/800, G-LOSM and Vampire 109 and 209 all took to the air for some local flying at Bournemouth. Hunter T.7 XL600 arrived by road on 25 June for rebuild. The aircraft had originally been surveyed by Eric back in December 1993, when offered for sale at £47,500. Initially registered as G-BVWN by a Southall owner at the end of 1994, it remained in storage. Purchased by Jet Heritage for £19,000 in June 1996, it was registered G-VETA in July on behalf of Mike Giles and a syndicate of Cathay Pacific captains who each owned a 10% share. Restoration commenced in the autumn, including the fitting of a new Avon engine. Airframe records showed that two of the Hunter T.7's worked on by Jet Heritage had suffered similar fatal incidents whilst in RAF service. XL572 (G-HNTR) had entered a severe spin on 27 August 1959 with both pilots attempting to eject. One did so, but was fatally injured. The second failed to leave the aircraft, which then recovered from its spin, enabling the shocked pilot to make a safe landing. XL600 (G-VETA) was undertaking aerobatics on 10 April 1963 when the second pilot fell out of the inverted aircraft as his seat was not locked. Again the accident resulted in a death.

In June Eric travelled to Airwork at Perth to inspect Hunter T.7 XX467 which had been in use as an instructional aircraft. Although in good order, the prospective purchaser didn't really know what he wanted to do with it, and so nothing further happened. Then in August Eric visited Marham to inspect some surplus Avon engines that were available. In order to get restoration work moving on the Swift, Eric appealed for sponsorship of £50,000, but none was forthcoming. He was also still on the lookout for an Attacker, as there were rumours that some still existed in Pakistan. Again nothing was forthcoming, despite King Hussein having made contact with the Pakistan Air Force on Eric's behalf.

Hunter One and Jet Heritage aircraft frequently visited European air shows. Seen at the Belgian Air Force St Truiden Show in September 1996, G-LOSM flies in company with a Spitfire and Hunter to represent three fighters flown by the Belgian AF. (MHP)

Bournemouth Airport held Big Nose Day on 21 April 1996 to mark the opening by Concorde G-BOAF of the runway extension. However, instead of having a fly-past of local aircraft, the airport had a 'tow-past' so as to save time. This involved aircraft from the various firms based at the airport, including Jet Heritage's Hunter XE677 and Vampire 209. The appearance of Concorde brought thousands of people to the airport, some of whom took the opportunity to visit Hangar 600 to see Jet Heritage's aircraft. G-LOSM appeared at North Weald's Fighter Meet on 11 May, but there were not so many display appearances as in the previous year. On 7 September there was a further commemorative flight relating to Neville Duke's speed record. During the summer Rustington Parish Council had erected a memorial to mark one end of the Sussex seafront course. Neville unveiled the memorial on the afternoon of 7 September and the event was overflown by XE677 piloted by Brian Henwood in company with Meteor G-LOSM. The following day G-LOSM appeared at St Truiden for the Belgian Air Force Show marking the fiftieth anniversary of the base, where it flew in formation with an OFMC Spitfire and Hunter. This was to commemorate three post-war fighters that had served with the Belgians. On the morning of 15 November there was a further Sir Frank Whittle commemorative fly-past by G-LOSM over Cranwell.

14

New Era at Jet Heritage

1997 brought major alterations to Jet Heritage's running, with a complete change in the management team. This saw the development of syndicate-owned aircraft, as well as plans for an expanded display fleet. When Adrian left, Ian and Dougal had become directors alongside Eric as a short-term measure only. It was their wish that someone else be found to run the company. It was also realised that Eric would be retiring as Chief Engineer somewhen in 1997 – when he reached 71! - so a replacement was sought. The formation of the charitable foundation and the Jordanian contract had not meant an end to the financial problems of running Jet Heritage. Additional income sources were still needed.

Ronan Harvey of Delta Jets at Kemble showed an interest in buying or investing in Jet Heritage, but discussions held in January 1997 came to nothing. However, there was interest elsewhere. On 6 February Mike Giles and Jonathon Whaley took over the company and the charitable foundation from Ian and Dougal. Jonathon was the new Joint Managing Director and Chief Pilot, a former FAA Sea Vixen and Phantom pilot, having 120 types in his log book after much competition, display and film work. Mike came on board as Joint Managing and Operations Director. He had flown Canberra PR.7s in the RAF and was currently a senior 747 captain with Cathay Pacific. Bernie Cox - former Chief Engineering Inspector at BAe Dunsfold - would be the new Director of Engineering from June. This meant there was little work for Eric, which in a sense was convenient as he had to spend a few weeks in hospital for a knee replacement. Long-serving secretary Carol James departed at the end of March – being redundant in the new set-up. Joining the team for a while as Chief Pilot was Dan Griffith, a former RAF Harrier pilot and test pilot at Bedford and Boscombe Down. On the engineering side, under the guidance of Bernie Cox, the mission was to keep the aircraft operational and to be the leader in historic jet engineering. On the charity side, the aims were to find new trustees and to build a museum to house a national collection of operational historic jet aircraft. Initially the collection would be housed in Hangar 600, but by 2000 it was planned that a more permanent site would be established on the northern side of the airport. Hangar 600 was a two-bay hangar – one to be for the museum, the other for engineering. To fund this ambitious new project details were

Although it was hard for him to realise, Eric was well past retirement age in 1997. Amongst the management changes within Jet Heritage at that time, a new Director of Engineering was appointed, which enabled Eric to retire on 1 August. (MHP)

announced in April 1997, which included that application was to be made for a £2m National Lottery grant. Hangar 332, adjacent to Hangar 600, had been used in recent months by Short Bros, but their lease expired early in 1997. Jet Heritage secured an option on the lease, but did not take it up as it would have been for a short duration only. Bournemouth Airport's master plan indicated that the site would be required sometime after 2000 for a new terminal building. For the same reason, Jet Heritage was unable to purchase the freehold of Hangar 600 – another reason to relocate to a new complex.

> *Jonathon:* The Museum project was a labour of love for most of those involved. I lived, ate, breathed and slept aircraft, and joined Jet Heritage because I wanted the public to be able to see these fantastic machines in action. We had the most important part – the aircraft – and needed fellow plane enthusiasts on board to enable the project to get up and running. When completed, it was intended to be the largest flying museum of jet aircraft in Britain.

The planned museum opening was put back to enable a more professional look to be given to the hangar and exhibits. For example, the intended entrance lobby and shop was still filled with rubber fuel tanks! Also, a viewing area had to be erected in the engineering hangar so that visitors could safely observe the work in progress. To cope with much of this work a band of volunteers was formed in May 1997, the first one being local enthusiast Richard Edwards.

Two new directors were appointed to Jet Heritage in February 1997. Jonathon Whaley (above) was Joint MD and Chief Pilot, with Mike Giles (left) as Joint MD and Operations Director. Both had fast jet experience – Jonathon in the Navy and Mike in the RAF. (JHL collection)

Richard: My RAF background was on radios and satellite communications, not aircraft. Having been back in civvy life for many years, I read the press item in April about the setting up of a museum. So I turned up the following day and was 'blown way' by the hangar full of Hunters and Vampires. I offered my services for free to Jonathon (anything to be with these aircraft) and so became the first volunteer. After a week Mike offered me the job of Special Projects Engineer on a one year contract. The grand title really meant I had become co-ordinator for the other volunteers, overseeing the jobs that needed doing around the hangar. It was soon obvious that the planned 1 July opening date could not be met and a year's postponement was agreed upon. This enabled me to oversee my first major project – the repainting of the interior of the hangar. For this protracted task over the autumn months the volunteers used 30 gallons of paint to cover 16,000 sq ft.

Early in 1997 BBC TV crews arrived at Jet Heritage with Robbie Coltrane to film a part of their *Men and Machines* series, entitled 'The Jet Engine'. Robbie explained the work of Frank Whittle and the subsequent development of jet engines in Britain. The practical working of an engine was demonstrated by Bob Tarrant, who ran up a Derwent on a test stand. Robbie then flew in the back seat of G-LOSM, piloted by Brian Henwood. In the spring the active Jet Heritage fleet was Hunter XE677 and *Heritage Pair* G-LOSM and 215, with the Swift still awaiting restoration. A new arrival was Hunter J-4104/G-PSST, which flew in from Dunsfold on 6 March for Jonathon's company – Heritage Aviation Developments Ltd. It had been stored at Dunsfold for another owner since its arrival from Switzerland the previous November and then flown down by Jonathon for overhaul and fitting of an electric start system. Having also become a 10% owner of G-VETA, Jonathon first flew it after restoration on 20 March, still unpainted at the time. Granted its Permit to Fly on 17 April, and after further flight trials, G-VETA was painted in a striking midnight blue colour scheme with gold trim. It was frequently flown during the summer and autumn months by its consortium of pilots under the title of TFB Aviation. Jet Heritage were represented at North Weald's Fighter Meet on 10 May by G-LOSM, G-VETA and XE677. G-LOSM then appeared at Duxford's Air Day on 8 June and at Ursel, Belgium on 13 July.

During the spring of 1997 work on the Jordanian Hunters and Vampires was coming to an end and arrangements were made to deliver the fleet to Jordan. This resulted in Eric, Richard Verrall and Brian Henwood all visiting Marka Air Base, Amman, in March. Hunters 800 and 843 departed early afternoon on 27 May with Vampires 109 and 209 following on 2 June. Hunter 712 remained with Jet Heritage at the request of King Hussein so that it could represent Jordan at air displays, undertaking limited work during the summer of 1997, including RIAT Fairford in July (it was flown to Delta Jets at Kemble in the spring of 1999, then delivered to Amman later that summer). Hunter XG160/G-BWAF was still in storage at the time. Its association with the famous 111 Squadron had been noted,

The first aircraft to appear from overhaul, following the arrival of the new management, was Hunter T.7 G-VETA. As usual, the Hunter undertook its initial flights in bare metal, having been flown first by Jonathon in March 1997. (MHP)

Hunter G-VETA was soon sprayed in a smart midnight blue colour scheme with gold trim. It was operated by a consortium of pilots who each contributed an equal share towards the operating costs which, not unexpectedly, were quite substantial. (MHP)

After delivery of four of the Jordanian Historic Flight aircraft in 1997, Hunter G-BWKC/712 remained with Jet Heritage in order to take part in UK air displays. This was at the request of King Hussein, who wished it to promote the Royal Jordanian Air Force. (MHP)

The Jet Heritage engineering team that worked with Eric at the time of his retirement in 1997. Paul Smith, Dave Thomas, John Gale, Bill Bates and Bob Tarrant. Behind them are some of the RJAFHF fleet. (Peter R March)

resulting in Richard Verrall requesting the MoD that it be loaned to Jet Heritage Charitable Foundation. The request was backed by King Hussein, resulting in the MoD's confirmation of the request in December 1997. The intention was to restore it to 111 Squadron colours as soon as time permitted.

Eric officially retired from Jet Heritage on 1 August 1997, feeling somewhat unhappy that he did not receive any thanks from the management team for the hard work he had put in over the last eight years. However his efforts were recognised by the aviation press. Peter March wrote in *Aircraft Illustrated* 'The team has been most fortunate in having Eric Hayward lead them through the restoration of twelve military jets'.

Eric: Part of the Jordanian contract had been for an engineer to visit Amman after a few months to see how things were going. As I was no longer working for Jet Heritage I did not want to be involved, but the Jordanians insisted I was the one they wanted. So I agreed to go to Marka for a month in October to check out how their engineers were coping. While there I spotted King Hussein, who happened to be visiting my hotel for a meeting. He came and shook my hand and asked how things were going. I explained that it was too early to tell. I knew the king was not in good health and his deterioration since I last saw him could be seen. Sadly he died in February 1999 – the passing of a good friend.

The Historic Flight was installed in its own hangar, but maintenance was not going too well. I had anticipated a dedicated team of engineers, whereas the Air Force kept changing them round. So there was no continuity and I found myself having to start from scratch with new engineers, sometimes on a daily basis. This

One last task for Eric was to visit the Jordanian Historic Flight at Marka in October 1997, to offer advice to their engineers. Seen with notepad in hand, Eric poses in front of Hunter 800 with the Flight's Chief Engineer. (EGH collection)

Whilst at Marka, the Historic Flight provided a 'demonstration' that Eric did not want to see. Vampire 109 undertook a test-flight and suffered loss of brakes on landing. So it was the classic 'wheels-up' landing to bring the aircraft to a halt, with the result seen here. (EGH collection)

Although in Royal Navy FRADU colours, Hunter WV372 was a T.7 not a T.8. Originally taken to North Weald for overhaul, the Hunter flew down to Bournemouth in August 1997 for overhaul by Jet Heritage on behalf of a consortium of pilots. (MHP)

restricted flying or else lead to unnecessary incidents. On 13 October I witnessed a test flight by Vampire 109 which, on landing, found it had no brakes due to the accumulator not being charged before the flight. So the pilot correctly went for the good old fashioned option – raise the undercarriage! It left the runway in a cloud of dust, resulting in the loss of its underwing tanks and damage to the under fuselage. It was removed to the hangar, but I had returned home before any decision had been made concerning repairs. Not a happy month. It also marked my final connection with Jet Heritage.

For the nearby Swanage Air Show on 2 August 1997, Jet Heritage provided Hunter XE677 and the *Heritage Pair.* The show was a headache for its organisers, who said it was a 'Free Show'. They intended to get their income from car parking charges, but the British public made other arrangements and walked. After a survey at North Weald by Jet Heritage's engineers, former RN FRADU Hunter T.7 WV372 arrived by air on 11 August for overhaul on behalf of a consortium, being registered G-BXFI. Another arrival from North Weald was Jet Provost T.5 G-BWOF on 25 October, which was hangared and maintained on behalf of Philip Meeson, joining his existing Dragon Rapide. During July further attempts were made to acquire the necessary parts for XR537/ G-NATY. Requests were even made in the USA, but all to no avail. So, regretfully, Frank Hackett-Jones agreed once more that work be suspended. It was decided to dispose of surplus Gnat XM697/G-NAAT as it was no longer required for spares. In its dilapidated state it was not a suitable aircraft for the proposed museum, just taking up storage space. Initially it was stored elsewhere on the airport, before departing by road to Dunsfold in July 1998. Across the runway the Jet Heritage engineers were impressed to see Source Classic Jets put up a formation of four Vampires and four Venoms for a photo shoot over the Isle of Wight in September 1997.

In his capacity as Airport Historian, the author checks the progress of Gnat XR537 after another attempt in the summer of 1997 to get it back into the air. Sourcing the required parts still proved a stumbling block – it was another ten years before XR537 flew. (Bournemouth Echo)

Having had one half of the hangar complex tidied up by the volunteers, it was then the turn of the engineering section to have a clear out to provide more working space.

Richard: Having learnt some organisational skills from running my own business, I realised that the volunteers needed some perks as a reward for all the hours they were putting in. As well as painting, they had spent their time getting rid of mountains of junk around the site. It was agreed that the one who had put in the most number of hours during the year would be rewarded with a flight in G-VETA. This turned out to be a retired group captain who had many Hunter hours under his belt from flying in the 1950s. So, the offer was passed to the next volunteer who willingly undertook a flight with Jonathon. On his return, the grin on his face filled the cockpit! Other volunteers also benefited from a flight in Philip Meeson's Dragon Rapide.

Another event I recall relates to when Bob Tarrant had to be sent to Amman to undertake further engineering work on the Hunters. For some reason he had to take a large amount of dollars with him, which was against regulations at the time. So, it was decided to equip him with a Hunter Servicing Manual, with the dollar bills hidden between the pages, which were then laminated together! This was not all of Bob's problems. Being an engineer he carried with him the usual tool box. Unfortunately, on

Jonathon and Mike were able to give their friends 'jollies' in one of the two-seat Hunters that were available. On a few occasions, some of the volunteers were able to experience such flights – hence the wide grin on the face of Julian Humphries, seen here with Jonathon. (JHL collection)

his return to the UK the Customs people examined the box, revealing a knife that Bob always carried with him – a tool of the trade. After much questioning, they retained the knife but let Bob go. He vowed never to return to Amman again!

Mike Giles had already headed up a syndicate of pilots, each owning a 10% share in G-VETA to cover its future operation. The directors decided to develop this idea further, with interest already shown in Hunter G-BXFI and a Jet Provost. In the spring of 1998 adverts were placed for syndicate partners, aimed at PPL pilots who wished to attain jet experience 'beyond their dreams'. The intention was to form syndicates of ten people (which was soon achieved for G-BXFI), who would then form an operating company. If required, Mike or Jonathon would assist by becoming a director. For the Jet Provost it would involve members contributing £8,000 each towards its purchase and restoration costs, plus an annual fee of £1,400 for hangarage, servicing and insurance. Then there would be flight training, fuel, landing fees – the costs kept mounting. There were plans to acquire former Swiss Vampire T.55 U-1222 which was for sale at SFr 80,000. Again a syndicate of ten was planned, this time contributing £6,000 each, but in the event the purchase did not proceed.

In the spring of 1998 Mike took over responsibility for the finances and running of the charitable foundation. This involved preparing the lottery grant bid for the

Another restoration project was Jet Provost 8458M, which arrived from the Isle of Man in the spring of 1998. The cross on the rear fuselage was to indicate to the Russians that the aircraft was no longer on effective RAF charge. (MHP)

new museum. Jet Provost T.4 8458M (G-RAFI) arrived by road in mid-March from Jurby, Isle of Man, where some limited restoration had been undertaken. The aircraft was purchased for £37,000 for syndicate operation, with restoration commencing in September. As well as working on the Jet Heritage aircraft, the new directors started clearing out what they regarded as unwanted equipment and accessories which had been acquired over the years. This was to provide the company with additional working capital. A return arrival on 18 April was John Davies' DH Venom G-GONE. This had been maintained by Jet Heritage between 1987 and 1995 and then placed in storage at Hawarden. It returned to Bournemouth for a major overhaul. Piloted by Jonathon, Hunter G-PSST first flew again after overhaul in May 1998, initially flying unpainted. Jet Heritage also had, or would soon have, access to eleven jets for air show displays. The planned star formations were the existing *Heritage Pair* Meteor G-LOSM & Vampire G-HELV/215 and the new *Hunter Four* (XE677/ G-HHUN, G-PSST, G-VETA and G-BXFI). Jet Provost G-BWOF was included, and it was intended to return Hunter XG160 and Gnat XR537 to the airworthy fleet, plus acquire an additional Vampire FB.6 and T.55 later in the year.

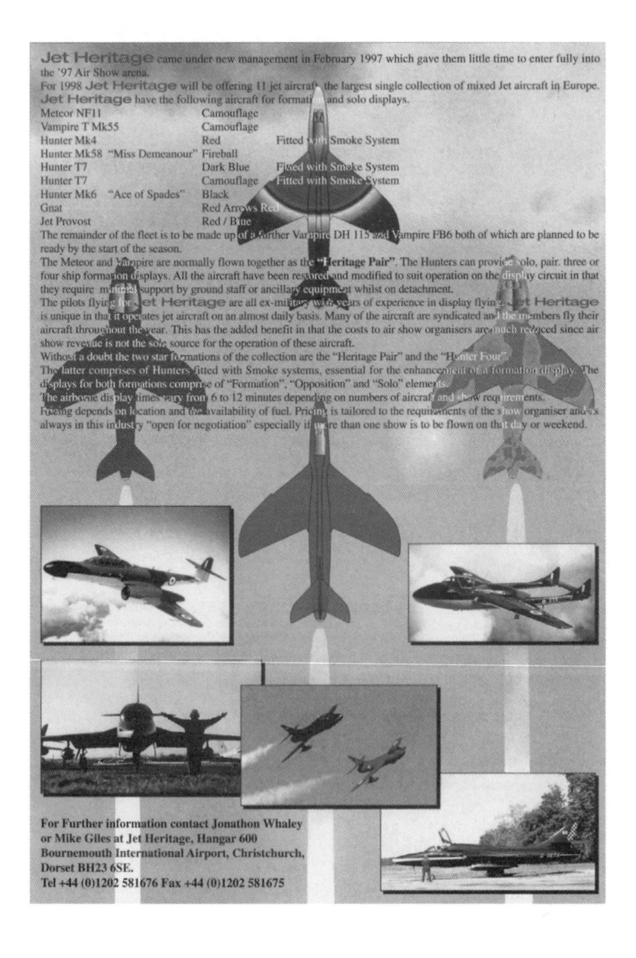

15

Museum at Last

Now coming to the fore, the Jet Heritage Charitable Foundation declared itself as the 'Guardian of aviation heritage'. Its long-term aim was finally realised on 17 May 1998, when the Jet Heritage Aviation Museum was opened to the public at 10.00 a.m., the event coinciding with the arrival of Concorde G-BOAG at the airport. The museum had always been Jet Heritage's intention, with plans originally being made back in 1983, but there had been problems to resolve over mixing the public with active jet aircraft. To overcome this, during the previous twelve months the team of volunteers had cleaned up the whole complex, repainted the hangar interior and erected unobtrusive barriers around the aircraft. An outside picnic area was provided which meant that there were good viewing opportunities when the aircraft were flying. For the opening day there were displays by Hunters XE677, G-VETA & 712, Jet Provost G-BWOF, Meteor G-LOSM and Vampire 215 shortly after Concorde had departed to Paris. There were additional displays in the afternoon, including a superb show by XE677, plus a more sedate performance by Philip Meeson's Dragon Rapide. At the time The Royal Saudi Air Force was showing interest in setting up a flying museum, along the same lines as Jordan. The display was visited by one of their Princes – Sheik Sultan bin Saeed Al-Qasimi - but there was no further progress. Plans were drawn up for the proposed hangar complex on the northern side of the airport. Worked on by Jonathon, Mike and Richard, as well as a museum it would be a centre of excellence. Based on a main hangar, there would be smaller ones around it in which other owners could work on their aircraft. Having aircraft on one site would hopefully mean that owners could assist each other with their projects. Unfortunately it turned out that time did not permit fund raising for the development of these plans.

A group of Swiss Hunter enthusiasts from Interlaken – the HUVER - visited on 24 May 1998. They were shown round the museum and given a short flying display by Mike Giles in G-VETA. Back in Switzerland, the Group had two Hunters of their own to look after. For RIAT Fairford on 25/26 July, the Heritage Pair were flown with other historic jets, including four Source Vampires, to mark the 80th Anniversary of the Royal Air Force.

A number of the fleet departed lunch time on 5 June to take part in that weekends Biggin Hill Air Fair (XE677, G-VETA and *Heritage Pair*). It was also BAe Families Day at Dunsfold and after practising its display routine in the early afternoon,

The opening of the Jet Heritage Museum in May 1998 included a flying display by the resident aircraft. This is the group of pilots who flew the fighters, with Mike Giles centre with cap. The press publicity helped to raise the profile of the museum. (MHP)

The setting up of the Jet Heritage Museum would not have been possible without help from the band of willing volunteers, led by Richard Edwards (shirtsleeves). They spent months painting and cleaning the hangars to make them presentable to the public. (JHL collection)

Jet Heritage Museum logo.

Shortly after the museum opening, a group of Hunter enthusiasts from Interlaken – the Huver – visited the complex to see the resident Hunters. G-VETA was displayed by Mike Giles (standing by cockpit) with new Director Glenn Lacey (in front of Mike) on hand. (JHL collection)

Hunter XE677 suffered a massive engine failure at low level and crashed onto the airfield, resulting in the death of its pilot John Davies. No attempt was made to eject from the stricken aircraft. The AAIB Accident Report commented that 'the pilot was in the command of a prestige historic jet aircraft which had been painstakingly restored and he may thus have felt that he should endeavour to recover the aircraft onto the runway'. A tragic day for historic aviation.

Having achieved the aim of opening a museum, all was not well at management level. Despite his enthusiasm for the aims of Jet Heritage, Jonathon Whaley was feeling the heat, having found that the set up was more traumatic than anticipated! The company was not going in the direction he anticipated and so he decided it was time to move on, officially departing on 1 August 1998. Jonathon took with him the long term Swift restoration project, which was regarded as too much of a drain time wise by the new management, departing by road to Scampton on 23 September. At that time a new owner for Jet Heritage was announced - Glenn Lacey - a businessman, who flew a company Cessna 421 twin and Auster 5 G-AKWS. This was already based at Bournemouth, having been painted in camouflage colours as RT610 during July. As such it was hangared with the collection, but obviously not part of it. Following Jonathon, Bernie Cox also departed, again having found that operations were not as expected. This resulted in the appointment of Paul Kingsbury to look after the engineering side. A former RAF Chief Engineer, Paul had experience on a wide variety of RAF aircraft and had already been with Jet Heritage for a few months. Mike Giles found that his '747 day job' kept him too busy elsewhere and so he

The state of G-SWIF after nine years with Jet Heritage. Overhaul work on the Jordanian and private-owner Hunters prevented the engineers from having sufficient time to work on the Swift, which turned out to be a more complex project than originally anticipated. (MHP)

stood down as Director. All these changes naturally had an unsettling effect. Also affecting operations was the fact that with an economic downturn in the country, investors and owners were no longer seeing a return on their investment. For example, the value of a Hunter had dropped from £220,000 to only £80,000.

There was an historic non-jet visitor to the museum in the middle of June 1998. Preserved C-121 Constellation N494TW crossed the Atlantic to take part in events marking the 50th Anniversary of the Berlin Air Lift. Painted in USAF colours, it stopped off at Bournemouth for a week before proceeding to Berlin. Naturally it attracted a number of visitors, although the owners were unable to give flights in the Constellation as they had originally hoped. In the early hours of 15 July Mig 21PF 503 arrived by road from North Weald for static display at the museum. Owned by the Mig 21 Group, and in Russian Air Force colours, its members undertook its external refurbishment over the next few months. After restoration Hunter G-BXFI was repainted during August into its original RAF 2 Squadron camouflage colours as WV372. As such it was flown again by Mike on 11 September on behalf of its new operators - the Fox One Group. Much flying was undertaken by G-VETA and WV372 giving conversion training to many pilots, plus a few 'jollies'. Early in the summer Karl Theurer was cleared to fly a Hunter. This meant he could now fly G-HVIP solo, not having to rely on Dan Griffith for instruction. In addition to the trademark Red Hunter badge of Jet Heritage, a Squadron Badge was introduced in September, still incorporating the Hunter and with the motto 'Preserve to Fly'. The Museum was now publicised as 'Jet Heritage Charitable Foundation Museum'.

Jet Heritage was represented by the *Heritage Pair* at Fairford's RIAT on 25/26 July, along with four of Source's Vampires. The *Heritage Pair* then displayed at Cranfield's Classic Jet Airshow on 15 August, with G-LOSM appearing at the

A surprise arrival in the summer of 1998 was this former Hungarian AF Mig 21, which had been stored at North Weald for some years. The group that owned the aircraft brought it to Bournemouth for restoration to static-display condition – there was no hope it would fly. (MHP)

Work under way in the engineering hangar in the autumn of 1998. In the foreground is Jet Provost 8458M, with Hunter J-4083 behind. This had been presented to Jet Heritage by the Swiss AF back in June 1995, but it was three years before it could be worked on. (JHL collection)

Proving to be the most noticeable Hunter in Europe was Jonathon's G-PSST. The aircraft has to be seen to fully appreciate the complex colour scheme which it received at the end of 1998. As normal for restored Hunters, it had been flying unpainted for a few months. (MHP)

Belgian AF's Kleine Brogel Show on 5 September followed by Duxford the next day. The Farnborough Air Show celebrated its 50th Anniversary in September 1998 and RJAF Hunter 712 was amongst the aircraft flying in the special display, with Source exhibiting six of its fleet in the static park. On 10 September the *Heritage Pair* took part in a special flight with the Meteor carrying the ashes of Sir Frank Whittle from Bournemouth, via Farnborough to RAF Cranwell, where the ashes were laid to rest at St Michael's Church. Sir Ken Hayr and Glenn Lacey were in the Meteor, with Mike Giles and Ian Whittle, Sir Frank's son, in the Vampire. On the afternoon of 9 November there was a further photo shoot by John Dibbs of the *Heritage Pair* and RJAF Hunter, using Strikemaster G-UNNY as camera ship. The autumn of 1998 saw restoration work commence on Jet Provost 8458M (G-RAFI) and Venom G-GONE. A further former RN Hunter T.8 was acquired - XF357/G-BWGL flying in from St Mawgan on 30 November still in its FRADU colours. However restoration did not commence until the following summer. A departure was WV372/G-BXFI which left for its new base at Kemble on 18 December. At the end of the year G-PSST emerged from the AIM paint shop in a striking colour scheme representing re-entry from space, the like of which hadn't been seen in the UK before. It has to be seen – words cannot describe it!

The new year of 1999 saw an additional new Chairman, John Hallett, in place alongside Glenn Lacey. John worked with the management at BAe Farnborough and had a Hunter, Sea Vixen and Vampire on display at Brooklands Museum. In April plans were put into place to split the organisation into an engineering section and museum side. Jet Heritage Ltd – the engineering side – would change its name to The Heritage Collection Ltd and continue with restoration as before. The Jet Heritage Charitable Foundation would continue to run the museum and handle fund raising. Limited restoration work commenced on the long stored ex-Swiss AF Hunter J-4083/G-EGHH early in 1999 and Jet Provost G-RAFI was painted in a grey colour scheme with 79 Squadron/1 TWU markings, flying again as XP672 on 27 February. Its overhaul had revealed that the airframe was not in as good a condition as thought, resulting in a ban being placed on it undertaking aerobatics. After delivery of their initial aircraft, the Royal Jordanian Air Force decided not to proceed with the overhaul of Hunter J-4081/G-BWKB. After a period of storage it was collected by OFMC and flown to their Scampton base on afternoon of 15 March. Hunter G-PSST was test flown during the spring, moving to Kemble in May to be maintained by Delta Jets. A further arrival at the museum in May was Channel Express' Handley Page Herald G-BEYF. This had been the last Herald flying in the world, having been withdrawn from service on 9 April and was presented on loan from Channel Express on 26 May. Although an airliner, it was fitting that the Herald be preserved at the museum due to the type's long association with Bournemouth Airport. An arrival in July was the fuselage of Canberra PR.7 WT532 which was donated by the airport to the museum having served with their fire section for some years. In reasonable condition, this gave the museum volunteers a restoration project to work on.

Although it was no a jet, the Jet Heritage Museum was happy to accept Herald G-BEYF for display. The Herald had strong associations with Bournemouth – the type's first service was from Jersey to Bournemouth (1961) and the last from Liverpool to Bournemouth (1999). (MHP)

By 1998 the Jordanian Historic Flight had decided that they would not require Hunter J-4081. After a period of storage at Bournemouth it was sold to the Old Flying Machine Company, and here prepares to depart to their Scampton base in March 1999. (MHP)

The weekend of 5/6 June was a busy one for Jet Heritage. As well as the final Jordanian Hunter 712 departing to Kemble before delivery to Amman, the *Heritage Pair* and XP672 made the regular appearance at Dunsfold's Family Day. The 6 June was also the museum's 1999 Open Day. As in the previous year the event saw British Airways Concorde G-BOAE operate from the airport. Following its departure displays were given by most of the airworthy aircraft, boosted by Canberra WK163 and Hunter G-PSST. Jet Provost XW433 was a visitors and Dakota G-DAKK undertook pleasure flights. The visiting public appreciated the benefit of Jet Heritage's viewing area, which was publicised as the place to be for Concorde visits (there were four visits in 1999). Further open days were planned – the next in September. The *Heritage Pair* Meteor and Vampire displayed at RIAT Fairford on 23-25 July, unknowingly at the time proving to be the final Jet Heritage display.

16
Disbandment

Despite the apparent optimism at the 1999 open day, all was not well financially within Jet Heritage. As from the summer it was planned that the company would continue under the new name of The Heritage Collection Ltd. One reason was that in the future it would overhaul historic piston aircraft as well as jets. However, Jet Heritage did not have time to change its name before it suddenly ceased trading, going into liquidation on 19 August. Little engineering work had been undertaken for anyone during the preceding weeks, although visitor numbers to the museum were increasing due to school holidays. The liquidation resulted in the CAA withdrawing its authority for the company to undertake flying and engineering. It was also found that, following the disposal of large quantities of spares and ground equipment, there were no assets left in Jet Heritage's name. As mentioned before, none of the aircraft were registered to Jet Heritage. As a result there were a number of creditors, including Bournemouth Airport and Esso, who came knocking on the hangar door.

By September Venom G-GONE and Hunter G-EGHH were moved across the airfield to the new firm of Phoenix Aviation formed by Glenn Lacey. The intention was to continue with their restoration, but the company also went out of business after a few months. Jet Provost XP672 was sold in September and flew away to North Weald. Hunter G-BWGL was re-assembled and flown out to Exeter for its restoration to be completed. Meteor G-LOSM and Vampire G-HELV remained in Hangar 600, whilst Hunter G-BWAF and dismantled Vampire G-SWIS continued in storage elsewhere on the airfield. However, G-LOSM was able to attend Duxford's Show on 12 September as arrangements had been made some months previously. It also flew on behalf of the RAF on 11 November. Marking the 50th Anniversary of Air-to-Air Refuelling, G-LOSM flew with a VC.10 tanker from Brize Norton over the Flight Refuelling site at Wimborne and then Bournemouth Airport. Fifty years before, a Meteor had successfully refuelled on a number of occasions from a Lancaster tanker during one flight, demonstrating the practicality of the new technique

To disassociate itself from the 'discredited' Jet Heritage name, the Jet Heritage Charitable Foundation was re-registered as the Bournemouth Aviation Charitable Foundation Ltd on the 27 August and continued in business as the Bournemouth Aviation Museum.

Following the liquidation of Jet Heritage, Venom G-GONE was left midway through its overhaul – hence the reason for its incomplete colour scheme. There was a delay of many months before it was able to return to the air. Hunter G-BWGL is in the background. (MHP)

Hunter G-BWGL had also been in Hangar 600 at the time of the Jet Heritage collapse. Overhaul work was almost complete, enabling the aircraft to be reassembled and cleared for a flight to Exeter for completion of its overhaul. (JHL collection)

Looking back, it was probably best that the name of Jet Heritage disappeared, although not in the manner that had been planned. The original aims of the founding directors were no longer relevant as the new directors intended to pursue different plans. Jet Heritage had always been associated with high-quality restoration work on former military fighters, especially Hunters. This was no longer the case.

So ended an era of historic jet fighter operations originally started by Mike Carlton back in September 1981.

Epilogue

Initially there was doom and gloom around Hangar 600 following the collapse of Jet Heritage. Thoughts were that the aircraft and experience would be lost – luckily this was not the case. Having run the museum side for fifteen months, the team of volunteers felt they had enough experience to continue. John Hallett had the same idea, giving his backing to the volunteers. The main problem appeared to be keeping the aircraft airworthy. However, this worry was soon eclipsed when it was realised the amount of debt that Jet Heritage had left behind, with hardly any assets left to pay the outstanding bills. It was back to the gloomy days of 1986.

Guided by an unpaid management team, the volunteers now ran the Bournemouth Aviation Museum, keeping its doors open to the public. Negotiations were held with the airport, the main creditor, over deferring the rent payments for a few months so that the Bournemouth Aviation Charitable Foundation could try and get up and running. New trustees came on board in December and this led the way to saving the museum. On the engineering side, De Havilland Aviation set up operations in the spring of 2000, taking on some of the former Jet Heritage engineers. The firm was approved for jet aircraft maintenance, getting G-LOSM, G-HELV and G-HVIP back into the air by the early summer. In their own right, they flew in Sea Vixen XP924 from Swansea in May 2000, the fighter becoming their flagship. In 2003 De Havilland produced a publicity brochure promoting their Jet Heritage services.

Museum operations happily carried on until the end of 2007, when a bombshell hit the museum. Bournemouth Airport was unwilling to renew the lease of the hangar after 31 December, as the area was needed for future car parking. At the time, airline services were expanding rapidly, and all land near the terminal was required for passenger needs. So, the museum closed its doors on 19 December, although De Havilland were able to stay in their half of Hangar 600 for a few more months, before moving to a new location on the airport in September 2008. The airport was unable to find a satisfactory site for the museum, and it seemed that all the hard work of the volunteers had been wasted. Salvation came in the autumn of 2008, when the museum was able to re-open on a new site at Adventure Wonderland adjacent to the airport. Interestingly, it was still possible to see operational Hunters at Bournemouth Airport in 2008. ZZ190/91 were owned by Hawker Hunter Aviation and operated by FR Aviation as target presentation aircraft under a Royal Navy contract. This

enabled Eric Hayward to continue to indulge his nostalgia for the fighter, having first worked on it over fifty years before.

Of the 'founding' members of Jet Heritage, Adrian Gjertsen continued with his 'day job' of flying Boeing 757/767s out of Gatwick. Eric Hayward retired locally, ensuring that he could keep an eye on airport movements from his window. The Craig-Wood brothers eventually sold their aircraft, concentrating on their other business interests.

Luckily, the names of Hunter One and Jet Heritage continue to be remembered amongst aviation enthusiasts for the pioneering work they undertook in preserving historic British jet fighters. Like-minded people around the world have since perpetuated their work. A fitting tribute.

Appendix I

Summary of Brencham/Hunter One operations

Commenced operations 1983. Ceased August 1986. Total, 3½ years

The following aircraft were received in non-flying condition, overhauled and restored to UK CAA Permit to Fly requirements. All were flown on the display circuit both in the UK and Europe during 3½ years of operations, without major incident.

Aircraft		*Hours flown on receipt*	*Hours flown Hunter One*
G-HUNT	Hunter 51	3630	127
G-BOOM	Hunter 7	2661	87
G-PROV	Jet Provost	1101	81
G-JETP	Jet Provost	2487	19
G-LOSM	Meteor 11	1408	34
	Total hours		348

Additional major work carried out included:

Electric start system fitted to both Hunters
Extended wing leading-edges and tail-braking parachute fitted to G-HUNT
Larger on-board battery capacity fitted to G-LOSM
Smoke generating capability fitted to both Hunters
All aircraft externally paint stripped and repainted

Appendix II

Hunter One/Jet Heritage Fleet

DH VAMPIRE FB.6   G-BVPO/109

Built by F+W at Emmen, Switzerland, in spring 1951. Delivered to Swiss AF as J-1106 in summer; remained in service for almost forty years. Modified in 1978 as target presentation aircraft, painted with vivid orange and black colour scheme; served with ZFK.5 at Sion. Withdrawn from service in 1990 and stored. Sold at auction to Swiss owner and registered HB-RVO. Purchased by RV Aviation in spring 1994 on behalf of RJAFHF and delivered to Jet Heritage at Bournemouth on 30 June, still in Swiss AF markings. Registered G-BVPO on 11 July, overhauled during summer and re-flown on 22 December 1994 in Royal Jordanian AF camouflage colours as 109. Undertook display work in UK prior to being delivered to Jordan by air on 2 June 1997 in company with 209. Suffered wheels-up landing at Marka on 13 October 1997 and stored.

DH VAMPIRE FB.6   J-1149/G-SWIS

Built by F+W at Emmen, Switzerland, in 1951 and delivered to Swiss AF as J-1149 in winter. Remained in service almost forty years. Out of service by 1990 and stored at Sion, J-1149 was presented to Jet Heritage in January 1991. Delivered by air to Bournemouth in early May and registered as G-SWIS. Limited restoration revealed a number of modifications required that would prevent aircraft gaining British P to F. Initially stored in Jet Heritage hangar and then off site. Registration cancelled, April 1997; still in storage, August 1999.

DH VAMPIRE T.11   XD599

Built by de Havilland at Hatfield in 1954; in RAF service operated by 1 Squadron, RAF College Cranwell and CATCS Shawbury. After withdrawal, XD599 was sold to Gloucester Technical College in December 1970 for ground instruction at Staverton Airport. Purchased by Doug Arnold in November 1980 and moved to Blackbushe to join his fleet of warbirds. Stored until purchased by Brencham in 1984 and to Bournemouth in March. Held in storage pending restoration, sold in November 1985 as part of Brencham's fleet reduction, and then to Caernarfon Air Museum in spring 1986.

DH VAMPIRE T.11 XH328

Built by de Havilland at Hawarden in 1956 and first issued by RAF to No.60 Squadron at Tengah, North Borneo, as squadron hack, before returning to UK for storage. Issued to No.3 CAACU Exeter, XH328 was withdrawn from service early in 1971 and stored at Keevil. Then stored at various locations around the country for the next twenty years. Purchased by the Craig-Wood brothers in 1987, major parts of XH328 moved from Cranfield to Bournemouth in December 1990, with registration G-VMPR reserved. In due course this lapsed, being taken up by another Vampire in 1995. No restoration work was undertaken by Jet Heritage, the airframe being stored in a dismantled state until sold off.

DH VAMPIRE T.11 XK623/'G-VAMP'

Late production Vampire built by de Havilland at Hawarden in the summer of 1956. For most of its RAF service operated by 5 FTS Oakington. Stored at St Athan before being sold to Hawker Siddeley in December 1968 for possible refurbishment. This did not happen and XK623 was donated to Moston Technical College, Manchester, for ground instructional use. Purchased by Brencham in spring 1984 and moved to Bournemouth where it was painted with registration G-VAMP, which correctly belonged to a Hot Air Balloon. Held in storage pending restoration, sold in November 1985, along with XD599, to the Caernarfon Air Museum.

DH VAMPIRE T.55 G-BVLM/209

Built by F+W at Emmen, Switzerland, in winter 1958, and delivered to Swiss AF as U-1216 in March 1959. In service until late 1980s when stored at Sion. Donated by the Swiss to RAF Benevolent Fund and delivered to Boscombe Down on 7 June 1990. Allocated RAF serial ZH563, it remained in full Swiss AF colours. After period of storage sold to RV Aviation/RJAFHF in spring 1994 and registered G-BVLM on 6 April. Flown to Jet Heritage in April, it carried three identities at the same time – U-1216, ZH563 and G-BVLM. Overhauled and painted in Royal Jordanian AF camouflage colours as 209, flying as such on 8 July 1994. [Original 209 delivered in 7/55.] Undertook display work in UK prior to being delivered to Jordan by air on 2 June 1997 in company with 109. Undertook limited flying with RJAFHF from its Marka base.

DH VAMPIRE T.55 G-HELV/'215'

Built by F+W at Emmen, Switzerland, in autumn 1958 and delivered to Swiss AF as U-1215. In service for thirty years and, following withdrawal in June 1990, was stored at Sion. Sold by Swiss AF at Auction in March 1991, U-1215 was purchased by Jet Heritage and delivered to Bournemouth on 28 August. Registered as G-HELV the following month, the Vampire was initially flown in its Swiss AF colours. After overhaul in spring 1993 repainted in pseudo-RAF camouflage colours with serial '215'. As such it flew on the display circuit as one half of the *Heritage Pair* alongside Meteor WM167. Still based with Jet Heritage in August 1999. [Later became 'XJ771' and eventually sold to Air Atlantique Classic Fighter fleet.]

DH VENOM FB.50 G-GONE

Built by F + W at Emmen in spring 1955 and delivered to Swiss AF as J-1542 in August. Following service, stored at Dubendorf in early 1980s. Purchased by Philip Meeson in September 1984, registered G-GONE and flown to Cranfield for restoration during 1985/6. Painted in RN Admiral's Barge colours and flown to Bournemouth January 1987. Further work by Jet Heritage, flying again July 1987. As well as being flown by Philip at displays, it was also flown by John Davies. Eventually John purchased fifty per cent share of G-GONE, he then purchased Venom outright. Moved to Hawarden in November 1994, returning for overhaul by Jet Heritage in April 1998. This work was still to be completed at time of Jet Heritage's collapse.

FOLLAND GNAT T.1 XM697/G-NAAT

Pre-production Gnat Trainer built by Folland at Hamble in summer 1961. Never delivered to RAF but used as trials aircraft at Dunsfold and Boscombe Down. Placed into storage on completion of trials, XM697 was presented to No.1349 Squadron ATC Woking in the mid-1970s. Displayed outside their headquarters, eventually painted in *Red Arrows* colours. Purchased by Jet Heritage, it was registered G-NAAT in November 1989 and moved to Bournemouth the following month. Being a non-standard airframe it was stored as a spares ship. During 1991 XM697 was repainted in RAF air superiority grey, but remained in storage. Being a non-flyer, the Gnat was sold in the autumn of 1997, departing to Dunsfold the following July.

FOLLAND GNAT T.1 XR537/G-NATY

Built by Folland at Hamble in spring 1963. Delivered to RAF in June and initially operated by No.4 FTS at Valley; later with the Central Flying School. During its time with the CFS, XR537 was operated by the *Red Arrows* from 1976. Withdrawn from flying in September 1979, it served as an instructional airframe with No.2 S of TT at Cosford as 8642M. Maintained in ground running condition, still in *Red Arrows* colours, XR537 was purchased by Jet Heritage in March 1990 on behalf of Frank Hackett-Jones, and registered as G-NATY in June. Overhaul almost complete by summer of 1991, having reached the stage of engine runs. Problems in obtaining final replacement parts resulted in work being suspended and the aircraft remained as static exhibit in the museum. [Finally took to the air in summer of 2007].

GLOSTER METEOR T.7 VZ638/G-JETM

Built by Gloster at Hucclecote in spring 1949, issued by RAF to No.500 Squadron at West Malling, VZ638 had a varied operational career. Its service days ended at the College of Air Warfare at Manby in 1965; then stored at Kemble. To Southend Museum in January 1972. Purchased by Mike Carlton at auction in May 1983, moved to Eastleigh for short-term storage and registered G-JETM in August. Arrived at Bournemouth in October and partial restoration work undertaken; externally restored in RAF training colours by 1985. Further work saw these colours removed by 1987. Entered in Christie's auction of October 1987. Unsold at the time and purchased by Aces High in December, departing by road to their North Weald base in January 1988.

GLOSTER (AW) METEOR NF.11 WM167/G-LOSM

Standard Meteor night fighter built by Armstrong Whitworth at Bagington in spring 1952. Delivered to No.228 OCU in August, then short periods with Colerne and Leeming Station Flights. Returned to Armstrong Whitworth in January 1961 for conversion to target tug TT.20 version. Undertook trials at Boscombe Down from January 1962, ending up with Flight Refuelling at Tarrant Rushton in March 1964 where it undertook trials with various towed targets. Withdrawn in July 1975 and flown to Farnborough, sold to Doug Arnold the following month. Delivered to Blackbushe in December, work was undertaken to return WM167 to its night fighter condition. Then stored before being purchased by Mike Carlton in spring 1984 and flown to Bournemouth. Registered G-LOSM in June 1984 and restored in colours of No.141 Squadron. Frequently flown at displays around UK and Europe, it was entered in Christie's auction of October 1987, but remained unsold. One of the first aircraft of the new Jet Heritage

WM167/G-LOSM was one of a number of Meteor NF.11's converted into TT.20 target tugs. Used for many years as a trials aircraft, it is seen in the summer of 1966 whilst in use by Flight Refuelling from Tarrant Rushton. (MHP)

fleet, it flew again in August 1991. WM167 was usually displayed alongside Vampire Trainer '215' as one of the *Heritage Pair.* Still with Jet Heritage at time of liquidation. [To Air Atlantique Classic Fighter fleet in May 2004.]

HAWKER HUNTER F.4 XE677/G-HHUN

Built by Hawker Aircraft at Blackpool in spring 1955. Delivered to RAF as XE677 in July and issued to 4 Squadron at Jever, Germany, in September. Later served with 93 and 11 Squadrons, finally to No.229 OCU at Chivenor in July 1957. When RAF Hunter F.4s were replaced by F.6s during 1958, XE677 was stored at No.5 MU, Kemble. At some stage it was one of the few F.4s to be modified with the saw-tooth wing leading edge, a characteristic of the F.6. In April 1961 it was delivered to Hawker at Dunsfold for possible refurbishment. This did not take place, and XE677 was donated to Loughborough University in January 1962 for use as an instructional airframe. Surplus to requirements in 1989, it moved to East Kirkby for storage. Purchased by Jet Heritage and moved to Bournemouth September 1989, registered G-HHUN in October. After restoration it flew again on 21 January 1994 as G-HHUN before being repainted overall red and reverting to its military marks as XE677. As such it resembled Neville Duke's record-breaking WB188 of September 1953 and Hunter One's first aircraft – G-HUNT. Actively flown and displayed by Jet Heritage, until destroyed in fatal crash at Dunsfold on 5 June 1998.

Originally delivered to the RAF in July 1955, Hunter G-HHUN was based at Jever, Germany, as XE677. During 1956/57 it was one of 93 Squadron's aerobatic team aircraft, and is seen on a training sortie during the summer of 1956. (G Kipp)

HAWKER HUNTER F.6A XG160/G-BWAF

Produced by Armstrong Whitworth at Baginton, Coventry, under subcontract to Hawker. Built summer 1956 and delivered to RAF as XG160 in October, initially to 43 Squadron at Leuchars. Later passed to 111 Squadron at North Weald and was one of their famous *Black Arrows* aerobatic aircraft during 1958. Flown by Sqd. Ldr Roger Topp in the squadron's record-breaking twenty-two aircraft loop at Farnborough in September 1958. Later operated by 229 OCU and No.1 TWU. On retirement from flying duties in the late 1980s it became an instructional airframe with TMTS as Scampton with serial 8831M. By 1990 its 111 Squadron connection was recognised, and it was repainted in 111's black colour scheme. Put up for disposal by RAF at the end of 1994. Presented to the RJAFHF, XG160 moved by road to Jet Heritage in January 1995, registered G-BWAF the following month. Because of its famous past, King Hussein agreed in December 1997 that the Hunter would not become part of the Flight, but would pass to the Jet Heritage Charitable Foundation. Stored for many years, it was restored to its 111 Squadron colours during 2003 and placed on display at the museum.

HAWKER HUNTER T.7 WV372/G-BXFI

Standard production F.Mk.4 built by Hawker Aircraft at Kingston in June 1955. Delivered to RAF in September as WV372, initially serving with 222 Squadron at

Leuchars from January 1956. Squadron disbanded in November 1957 following defence cut-backs and WV372 was stored. Returned to Armstrong Whitworth at Bitteswell spring 1959 for conversion to T.Mk.7 trainer and re-delivered to RAF. Issued to RAF Germany in May 1959, returned to UK and then brief period with RAE Farnborough from March to July 1971. Stored again before being issued to 4 FTS at Valley in 1977, later returning to Germany. Then to 208 Squadron at Lossiemouth to train Buccaneer crews, later passing to No.237 OCU. Transferred to Royal Navy in October 1984 and issued to FRADU at Yeovilton. Even though operated by the navy it remained a T.7 – the other trainers in navy service were T.8s. Withdrawn early in 1993 and flown to Culdrose in March for ground use by SAH. Sold at auction, registered G-BXFI in April 1997 and flown to North Weald in May. Purchased by Jet Heritage July 1997 and flown to Bournemouth 11 August, still in RN colours, for overhaul. To Fox-One Group in April 1998, repainted in former 2 Squadron RAF colours, flown again 10 September and delivered to Kemble in November.

HAWKER HUNTER T.7 XL572/G-HNTR

Built by Hawker Aircraft at Kingston in spring 1958, XL572 was delivered to RAF in July and issued to No.229 OCU. Later served with both 2 TWU and 1 TWU before being withdrawn in 1984. To No.2 SoTT at Cosford in August 1984 with serial 8834M for ground instructional use. Sold in summer 1988 via Lovaux of Bournemouth and moved to Bournemouth in October for Jet Heritage. Registered G-HNTR in July 1989 in anticipation of restoration. In the event held in storage until sold to British Aerospace in October 1991 for ground instruction use and moved by road to their Overseas Training Department at Brough.

HAWKER HUNTER T.7 XL600/G-VETA

Built at Kingston and delivered to RAF as XL600 in November 1958. Issued to 65 Sqd. in December, later to 111 Sqd. and then various training units, including 4 FTS. To Scampton in December 1983 for ground instruction use with TMTS. Relinquished by RAF to RN in March 1985 for further period of instruction at Fleetlands as A2729. Sold to private owner at Southall in March 1992 and registered G-BVWN in December 1994. Remained in storage at Southall until purchased in June 1996 by group of Cathay Pacific pilots and moved to Bournemouth. Re-registered G-VETA in July and rebuilt by Jet Heritage, flying again March 1997. Sold to new owner at Kemble in November 2000.

During its RAF service Hunter XL600/G-VETA operated with many units; these included a spell as '83' with 4 FTS at Valley during the early 1970s. It is seen here at Odiham during Farnborough Air Show week in September 1970. (MHP)

HAWKER HUNTER T.7 XL617/G-HHNT

Built by Hawker Aircraft at Kingston in winter 1958. Delivered to RAF as XL617 in January 1959 and issued to 4 Squadron in Germany. Then passed to Jever and Gutersloh SF before returning to UK in 1964. Served with No.229 OCU and 1 TWU before being relegated to ground instruction use. To 2 SoTT at Cosford in September 1984 as 8837M. Sold in summer 1988 via Lovaux of Bournemouth and moved to Bournemouth in October for Jet Heritage. Registered G-HHNT in July 1989 pending restoration, but remained in storage. Sold autumn 1989 to Northern Lights of Canada and shipped by sea to USA in November. Restored to airworthy condition as N617NL.

HAWKER HUNTER T.7 G-BOOM/'800'

Built by Hawker at Dunsfold in summer 1958 and delivered to Dutch AF in October with serial N-307. Surplus to requirements by 1967 and sold to Danish AF in December with serial ET-274. Operated by Esk.724 at Aalborg, remained in service until November 1973. Repurchased by Hawker Siddeley in December 1975 and transported to Dunsfold for possible refurbishment. Remained unused, ET-274 moved to Hatfield for further storage. Sold to Brian Kay in June 1979, moving to Leavesden for rebuild and registered G-BOOM in October 1980. Limited flying from Stansted in 1981/82, before being sold to Mike Carlton in September 1982. Repainted in red colour scheme and familiar at UK and European airshows, frequently in company with G-HUNT. Flagship aircraft of Hunter One, entered in Christie's auction of October 1987 but remained unsold.

G-BOOM became one of the original members of Jet Heritage and remained active on display circuit. Sold at auction November 1993, it then became founding member of RJAFHF in spring 1994, flying in June in Jordanian colours with serial 800. [The original 800 saw service in Jordan in the 1970s.] Initially remained with Jet Heritage to undertake display work in the UK. Finally delivered by air to RJAFHF in May 1997 and based at their Marka Amman HQ.

HAWKER HUNTER T.8C XF357/G-BWGL

Originally F.Mk.4 built by Hawker Aircraft at Blackpool in 1956. Delivered to RAF as XF357 in March and issued to 130 Squadron, Bruggen. Squadron disbanded in spring 1957 and XF357 returned to UK for storage. Released by RAF, converted by Armstrong Whitworth at Bitteswell into T.Mk.8 trainer for the Royal Navy and delivered in May 1959. Saw service with 738 Squadron at Brawdy until May 1970, then to FRADU at Yeovilton, then upgraded to T.Mk.8c. Withdrawn in 1994 and stored at Shawbury. Sold to private owner in July 1995 and registered G-BWGL. Initially delivered to Exeter before being flown to St Mawgan in 1998 for short-term storage. Purchased by Glenn Lacey in December 1998 and flown to Bournemouth in November, still in FRADU colours with code 871. Overhaul by Jet Heritage commenced early 1999, but not fully completed. Re-assembled in September as G-BWGL and flown to Exeter in November. [Later to OFMC at Scampton.]

HAWKER HUNTER F.51 G-HUNT

Initial flagship of Hunter One, built by Hawker Aircraft at Kingston in spring 1956. Delivered to Danish AF in June with serial E-418 and spent most of its service life with Esk.724 at Aalborg. Withdrawn in 1975 with 3,574 hours on the airframe and stored. Purchased by Hawker Siddeley in December 1975 for possible refurbishment and returned to Dunsfold with a number of others. Then purchased by Spencer Flack in spring 1978, by road to Elstree in May and registered G-HUNT. Major rebuild over next two years and flown again 20 March 1980. Undertook display work, sold to Mike Carlton September 1981 and flown to new base at Bournemouth. First aircraft of Hunter One collection. Undertook display work, frequently in conjunction with G-BOOM, until summer 1986. Following winding up of Hunter One, entered in Christie's auction of October 1987. Purchased in error by Irish dealer and resold December to Jim Robinson in USA. Delivered by sea re-registered N611JR and painted with RAF roundels and serial WB188 – so resembling Neville Duke's record-breaking Hunter of 1953. Used for display flying and racing, eventually being presented to Oshkosh Museum.

Seen departing Aalborg, Hunter E-418/G-HUNT was originally delivered to the Danish Air Force in the summer of 1956. By the end of its service life with Esk 724 thirty years later, it was wearing a much-worn drab green camouflage scheme. (H Larsen)

HAWKER HUNTER F.58 G-BWKA/'843'

One of the original batch of 100 Swiss AF Hunters, built by Hawker Aircraft at Kingston during autumn 1959. Delivered to Swiss and entered service with Fl.St.3 in December with serial J-4075. Saw extensive service, finally operated by Fl.Rgt.3 until November 1994 when stored at Emmen and put up for sale. Presented by Swiss AF to RJAFHF and delivered by air to Jet Heritage as part of four-ship formation on 16 June 1995. Registered G-BWKA in October and overhauled for RJAFHF. Flown again June 1996 in Jordanian AF colours with serial '843'. [Original 843 delivered autumn 1971]. Delivered by air to Jordan 27 May 1997 along with '800'. Later stored at Marka.

HAWKER HUNTER F.58 J-4081/G-BWKB

Built by Hawker Aircraft at Kingston in autumn 1959. Delivered to Swiss Air Force; entered service with Esc.Av.5 at Rarogne in December with serial J-4081 and saw extensive service. Operated by Esc.Av.5 until November 1994 when placed into storage. Donated by Swiss AF to RJAFHF and flown to Jet Heritage as part of four-ship formation on 16 June 1995. Registered G-BWKB in October but remained stored in Swiss colours. Eventually released by RJAFHF, sold to Old Flying Machine Co. in February 1998 and flown to their Scampton base in March 1999.

HAWKER HUNTER F.58 G-BWKC/'712'

Built by Hawker Aircraft at Kingston in spring 1959. Delivered to Swiss AF and entered service in April with serial J-4025; saw extensive service including Fl.St.20 and Patrouille Suisse aerobatic team. Withdrawn in December 1994 and stored at Emmen. Presented to RAF Benevolent Fund and flown to Fairford in July 1995 for display at RIAT still in Patrouille Suisse special 'thirty years' colours. J-4025 then passed to RJAFHF and flown to Bournemouth at the end of July for overhaul by Jet Heritage. Registered G-BWKC in October and after overhaul appeared in Jordanian colours with serial 712. Flown again August 1996 and remained in UK in order to represent Jordan at air shows. Moved to Kemble in spring 1999 before delivery to Marka later that summer.

HAWKER HUNTER F.58 J-4083/G-EGHH

Built by Hawker Aircraft at Kingston in autumn 1959 and delivered to Swiss AF, entering service in January 1960 with serial J-4083. Saw extensive service, including Esc.Av.5 and Fl.St.21 at Raron, before being withdrawn in September 1994 and stored at Interlaken. Presented to Jet Heritage and flown

The Swiss Air Force was a major operator of the Hunter for thirty-five years, when it was used in both the interception and ground attack role. J-4083/G-EGHH served in the latter role, being modified to carry Maverick missiles, finally serving with Fl St 21 at Raron. (Brian S Strickland)

to Bournemouth as part of four-ship formation on 16 June 1995. Registered G-EGHH in July 1995 but remained in storage whilst work undertaken on Jordanian aircraft. Only limited amount of work had been carried out by the time of the collapse of Jet Heritage in 1999. [Eventually sold and moved by road to Kemble.]

HAWKER HUNTER F.58A J-4104/G-PSST

Originally F.Mk.4 built by Hawker Aircraft at Blackpool in summer 1956. Delivered to RAF in April with serial XF947 and initially issued to 3 Squadron at Geilenkirchen, Germany. Saw service with No.229 OCU before being withdrawn in 1964 and stored. Transferred to Royal Navy in February 1965 for ground instruction use with serial A2568 at HMS *Condor*, Arbroath. Purchased by Hawker Siddeley for refurbishment at Dunsfold during winter 1971 as F.58A for Swiss AF. Delivered by road in January 1972 with serial J-4104 and delivered to Fl.St.11. Final operator was Fl.St.20 at Mollis, where it remained until December 1994. Stored at Emmen with other Swiss Hunters; sold in spring 1995 but remained in Switzerland. At end of 1996 flown to Dunsfold for further storage prior to being purchased by Jonathon Whaley's Heritage Aviation Development in January 1997. Registered G-PSST (representing Personal Super-Sonic Transport), delivered to Jet Heritage in March for rebuild, flying again in May 1998. Repainted in striking colour scheme at Bournemouth, re-appearing in January 1999. Flown to new base at Kemble in May.

HAWKER HUNTER T.68 J-4208/G-HVIP

Originally Hunter F.Mk.50 built by Hawker Aircraft at Blackpool late 1956 and delivered to Swedish AF in January 1957 with serial 34080. Served with F.18, F.10 and finally F.9. Withdrawn in January 1969 and placed into storage. Repurchased by Hawker Siddeley and flown to Dunsfold in 1972 for refurbishment. This involved it being converted into a T.Mk.68 for the Swiss AF and delivered by road in May 1975 with serial J-4208 *Friar Tuck*. Remained in Swiss service until 1993 when placed into storage. Presented to Golden Europe Jet Deluxe Club early in 1995 and delivered to Bournemouth as part of four-ship formation on 16 June 1995. Registered as G-HVIP by Golden Jet in July 1995 and overhauled by Jet Heritage. Owner lived in Germany, but German regulations prevented former military jets being operated on their civil register. Flown again in April 1996, G-HVIP remained housed with Jet Heritage on behalf of its owner, who frequently visited Bournemouth to fly his aircraft. [Moved to Swiss base in 2004 and sold to Northern Lights, Canada, in spring 2008.]

Hunter Trainer J-4208/G-HVIP was the final Hunter delivered to the Swiss Air Force. As such it carried special nose art, depicting Friar Tuck, and is seen on roll-out at Altenrhein in June 1975. The final Swiss fighter version depicted Robin Hood. (EGH Collection)

HAWKER HUNTER 'BHAC'

A hybrid airframe, the fuselage of which was a T.7 trainer built by Hawker Aircraft at Kingston in summer 1958. Delivered to Danish AF in December with serial ET-272, it mainly served with Esk.724 at Aalborg along with ET-274 (G-BOOM). Withdrawn by mid-1970s and purchased by Hawker Siddeley in December 1975 for possible refurbishment. Remained unconverted at Dunsfold and moved to Hatfield in March 1976. Hulk, without cockpit section, purchased by Spencer Flack to provide spares for rebuild of G-HUNT. Remains moved to Leavesden in 1981 to provide spares for G-BOOM. Sold again to Mike Carlton and moved to Bournemouth late 1982. After being stripped of any remaining useful parts, fuselage prepared for gate guard duties at Biggin Hill. Fitted with different wings, spare fighter nose and so, when viewed, does not appear to be a trainer. Mounted on pole outside Brencham HQ at Biggin Hill in May 1985 with registration 'BHAC' for Brencham Historic Aircraft Co. Following break-up of Hunter One, purchased by Jet Heritage and returned to Bournemouth in November 1989. Refurbished and placed on pole in June 1991 as gate guard for Jet Heritage, with memorial plaque dedicated to Mike Carlton. Remained in place even when museum moved in 2008.

HAWKER SEA HAWK FB.5 WM994/G-SEAH

Built as FB.Mk.3 by Armstrong Whitworth at Bagington in summer 1954. Delivered to Royal Navy in August with serial WM994, and served with 800, 764 and 767 Squadrons. During 1956 it was upgraded to FB.Mk.5 version. Then served with 806 Squadron and from 1958 at Hal Far, Malta. Withdrawn and stored at Abbotsinch, Scotland in November 1960. Then to RNAS Arbroath for ground instruction duties with serial A2503. Later passed to Cranfield College for further instructional use. Sold in mid-1970s and moved to Swansea for restoration; registered G-SEAH in April 1983. Moved to Southend the following month, but little further work was undertaken. Bought by Mike Carlton and moved by road to Bournemouth in October 1983. Restoration work commenced, but incomplete by time of Christie's auction in October 1987. Failed to sell, but bought by Adrian Gjertsen in January 1988 and registered to Sark International Airways. Remained with Jet Heritage, but Hunter workload meant little further restoration was undertaken. To Amjet in Minneapolis in November 1994 as N994WM in part-exchange for a Strikemaster, departing by sea.

HAWKER SEA HAWK FGA.6 WV795

Built as FGA.Mk.4 by Armstrong Whitworth at Bagington summer 1954. Delivered to Royal Navy with serial WV795 September 1954 and served with 700, 806 and 738 Squadrons. During this time upgraded to FGA.Mk.6 version. On withdrawal stored at RNAS at Sydenham. Transferred to RAF as instructional airframe in 1969 and delivered to Halton with serial 8151M. To RNAS Culdrose in March 1976 for further period of instructional use. Sold in summer 1978 to a Cardiff collector, had moved to Bath area by summer 1981 for restoration. Purchased by Jet Heritage in October 1989 and moved to Bournemouth, still as 8151M. Initially stored but cosmetically repainted as WV795 by spring 1992. Remained in store at Jet Heritage, even when purchased by Cypriot Museum owner in 1993. He lost interest in his purchase, and WV795 departed by road to Bruntingthorpe in August 1995.

HAWKER SEA HAWK FGA.6 XE489/G-JETH

Last but one Sea Hawk FGA.Mk.6 produced by Armstrong Whitworth at Bagington in winter 1955. Delivered to Royal Navy as XE489 in January 1956 and issued to 899 Squadron. Stored after front-line duties, and then to Airwork FRU at Bournemouth in autumn 1961. Out of service by early 1967, XE489 was presented to Southend Museum in May 1967. After delivery, it was repainted as XE364 in Suez marking. Purchased at Auction in May 1983 by Mike

After service with the Royal Navy, Sea Hawk 'XE364'/G-JETH was delivered to the Historic Aircraft Museum at Southend in May 1967, where it remained for sixteen years. Displayed in Suez Markings, it is seen at the auction of the museum's collection in May 1983. (MHP)

Carlton and initially moved to Eastleigh. Registered G-JETH in August and to Bournemouth in October 1983 still as 'XE364'. When restoration commenced it was found that inner-wing spares had been cut and new set of wings was sought. WM983 was borrowed from Sea Cadets at Chilton Cantelo in 1985 in order to swap wings. However, there was also to be a swap of identities. WM983 was not returned to the Sea Cadets, but repainted red with no marking to match G-HUNT. During the summer of 1986 it acted as gate guard to the Bournemouth Flying Club. Entered in the Christie's auction of October 1987 as G-JETH, it should correctly be described as G-JETH (2nd). XE489/G-JETH (1st) was repainted as WM983 (2nd) and returned to the Sea Cadets in March 1987. The Cadets sold it in August 1989 to the Dutch Museum at Soesterburg where it was initially displayed as 'Kon Marine 131'.

HUNTING PROVOST T.1 XF877/G-AWVF

The only non-jet of the Jet Heritage fleet. Basic trainer built by Percival at Luton in summer 1955 and delivered to RAF as XF877. Served with RAF College and CNCS before withdrawal in 1968. Sold to private owner as G-AWVF in November 1968. Purchased by Jet Heritage in December 1990 with intention of displaying alongside Jet Provosts in its original RAF colours. However, the airframe hours prevented XF877 being used for aerobatics and so it was stored, mainly at Thatcham and Goodwood, from summer of 1997. Eventually sold.

HUNTING JET PROVOST T.4 XP672/G-RAFI

Built by Hunting Aircraft at Luton in summer 1962, delivered to RAF with serial XP672 and served with various training units. These included No.2 FTS at Syerston, CofAW at Manby, CATCS at Shawbury, finally returning to CofAW again. Out of service by 1975, XP672 was transferred to No.1 SoTT at Halton as ground trainer in January 1976 with serial 8458M. Surplus to requirements in 1991, sold to private owner in Isle of Man and registered G-RAFI in December 1992. However, little restoration work was undertaken. Purchased by Jet Heritage in February 1998 and by road to Bournemouth on 26 March still as 8458M. To Glenn Lacey November, overhauled and repainted in colours of 1 TWU/79 Squadron by January 1999. Flown again in February, sold in summer following break-up of Jet Heritage and flown out to North Weald.

HUNTING JET PROVOST T.4 XR658

Built by Hunting at Luton in spring 1963 and delivered to RAF as XR658 and served with 7 FTS, CAW and 6 FTS. Crash-landed in October 1971 and transferred to Abingdon for ground instruction use as 8192M. On withdrawn stored at Wroughton, from where it was purchased by Jet Heritage in June 1988. As part of the Swift deal, it was delivered to Flint College in June 1990 for ground instruction use.

HUNTING JET PROVOST T.52 G-JETP

Originally built by BAC at Warton as T.Mk.4 in summer 1962. Delivered to RAF with serial XP666 and issued to No.7 FTS. Repurchased by BAC and returned to Warton in August 1967 for conversion to T.Mk.52 for South Arabian AF. Delivered to Aden in spring 1968 with serial 107. Withdrawn by late 1977 and sold to Singapore AF in March 1978 with serial 355. After only short period of use, withdrawn in August 1981 and stored at Tengah. Purchased by Brencham in August 1983 and shipped back to Bournemouth in November, registered G-JETP. Overhauled during spring 1985 and flown 19 July, later painted black overall. Limited flying until September 1986 and then held for Christie's auction of October 1987. Sold to Craig-Wood brother (the other purchased G-PROV) and became one of the founding aircraft of Jet Heritage. Part of the display team, it was sold at Sotheby's auction in September 1992 to a Cypriot collector. Delivered by air in May 1993, it remained inactive with its new owner.

HUNTING JET PROVOST T.52 G-PROV

Originally built by BAC at Warton as T.Mk.4 in spring 1964. Delivered to RAF with serial XS228 but placed into storage. Purchased by BAC and retuned to Warton in January 1967 for conversion to T.Mk.52 for South Arabian AF. Delivered to Aden in October 1967 with serial 104. Withdrawn by December 1972 and placed into storage. Sold to Singapore AF in February 1975 with serial 352. Remained in service until October 1981, when placed into storage at Tengah. Purchased by Brencham in April 1983 and shipped back to Bournemouth in November, registered G-PROV. Overhauled during summer 1984, first flying in November and then painted red overall to match G-HUNT. Undertook display work during next two years. Entered in Christie's auction of October 1987 and purchased by Craig-Wood brother (other buying G-JETP). Remained at Bournemouth and became founding aircraft of Jet Heritage. Sold in March 1993 and delivered by air to Leavesden.

NORTH AMERICAN F-86A SABRE 48-178/G-SABR

Early production F-86A fighter built by North American Aviation in autumn 1948. Delivered to USAF 1st Fighter Group at March AFB in April 1949. Being an early version, 48-178 did not serve long in front-line USAF service, ending its days with California ANG at Ontario. After serving as ground instruction airframe, withdrawn by August 1957 and sold for scrap in February 1958. Later rescued from scrap yard by private owner, registered as N68388 and slowly restored to airworthy condition. Re-flown in mid-1980s and registered NX178. Purchased by Golden Apple Organisation in November 1991 and registered G-SABR. Delivered by sea to Bournemouth in January 1992 and reassembled by Jet Heritage. Retained its USAF colours, first UK flight on 21 May. Undertook display work on behalf of Golden Apple, moving base to Duxford in March 1994.

SUPERMARINE SWIFT F.7 XF114/G-SWIF

One of twelve F.Mk.7s built by Vickers-Supermarine at South Marston in the summer of 1956. This version was used by the RAF for guided missile trials, but XF114 was retained by the MoD for other trials. These included being based at Filton for tests with Bristol Siddeley Engines. During 1962-5 it was based at Cranfield for braking trials on wet runway surfaces and after completion was stored at Aston Down. In April 1967 XF114 was delivered to the Flint Technical College in North Wales for ground instruction duties. Its historic value was spotted by Mike Carlton who negotiated to acquire the aircraft. Following

Used as a trials aircraft for many years, Swift XF114/G-SWIF acquired an overall black paint scheme during the time it was used for runway braking trials. It is seen whilst based at Cranfield in 1962, from where it also undertook trials at Heathrow. (Peter R March)

Mike's death, the purchase was delayed, finally being completed by Jet Heritage in November 1988. Moved to Bournemouth on 19 January 1989, it was a jewel in Jet Heritage's initial fleet, albeit requiring a lengthy rebuild. Registered G-SWIF in June 1990, restoration proceeded at a slow pace during the early 1990s, but was suspended due to pressure of work on RJAFHF Hunters. Project taken on by Jonathon Whaley when he left Jet Heritage and G-SWIF departed by road for Scampton in September 1998. [In the end thoughts of restoration to flying condition had to be given up and XF114 ended up with the Solent Sky Museum.]

Index

Aalborg Air Base 15
Amman 98, 119, 121
Arnold, Doug 46, 83, 84, 90

Biggin Hill 24, 38, 63, 110
Bournemouth Air Shows 47, 64, 74, 86, 115
Bournemouth Aviation Museum 135, 137
Bournemouth Helicopters 92
Bradshaw, John 59

CAA 19, 33, 82
Caen 45
Chilton Cantelo Sea Cadets 55
Christie's Auction 67
Classic Jets 110
Coltrane, Robbie 119
Cornah's Quay College 54
Constantinides, Saavas 90, 92
Constellation C- 121 130
Coventry 25, 27
Cranfield 20, 21
Cranwell 132

Davies, John 92, 129
De Havilland Aviation 137
Delta Jets 116
Display routine 49
Duke, Neville 11, 38, 52, 94, 115
Dunsfold 13, 17, 24, 36, 127

Eagle Beechcraft 64, 72
Eastleigh 40, 65
Elstree Air Force 13, 15, 24
Ernmen 103

Farnborough 132
Flight Refuelling 49, 135
Fox One Group 130
FR Aviation 42, 137
Glos-Air Services 29, 42, 57
Golden Apple Trust 89
Golden Europe Jet Club 108

Hackett-Jones, Frank 80, 109, 114, 123
Haydon-Baillie, Ormond 11
Heritage Collection 135
Home, Robert 80, 89
Huver 127

Jet Heritage Aerospace Division 86
Jet Heritage Aviation Museum 106, 127

Jet Heritage Charitable Foundation 95, 106, 127
Jet Heritage Film Service 86

Kay, Brian 21, 22, 33
King Hussein of Jordan 96, 103, 104
Klagenfixt 49

Leavesden 22
Liege 44
Little Staughton 31
London to Paris Air Race 3 8
Loughborough University 78
Lovaux 65, 66

Meeson, Philip 66, 77
Meteor 50th Anniversary 91
Metropolitan Airways 51, 59
Mildenhall 54
Moston College 43

New Milton ATC 80
North Weald 85

Ostend 54

Patterson House 35, 54, 65
Phoenix Aviation 135

Return to Fairborough 51
RIAT Fairford 104, 127, 134
Robinson, Jim 71
Rolls-Royce 81, 82
RV Aviation 96

Singapore 33
Sion 84, 89
Sotheby's Auction 80, 90, 95
Southend Museum 40
Source Classic Jets 94, 123
Super-X Simulators 77
Swanage Air Show 123
Swiss Air Force 84, 89, 103, 111

The Black Knights 12
Theurer, Dr Karl 108, 114, 130

Valley 25, 31
Vampire 50th Anniversary 94
Verrall, Richard 96

Waddington 25
Whittle, Sir Frank 84, 132
Wrecks & Relics 20